Nanotoxicology in Nanobiomedicine

P K Gupta

Nanotoxicology in Nanobiomedicine

 Springer

P K Gupta
Bareilly, Uttar Pradesh, India

ISBN 978-3-031-24289-2   ISBN 978-3-031-24287-8   (eBook)
https://doi.org/10.1007/978-3-031-24287-8

This Springer imprint is published by the registered company Springer Nature Switzerland AG
The registered company address is: Gewerbestrasse 11, 6330 Cham, Switzerland

10,000 years ago, man domesticated plants and animals, now it's time to domesticate molecules.

– Professor Susan Lindquist

Susan Lee Lindquist *(June 5, 1949–October 27, 2016) was an American professor of biology at MIT specializing in molecular biology, particularly the protein folding problem within a family of molecules known as heat-shock proteins, and prions. Lindquist was a member and former director of the Whitehead Institute and was awarded the National Medal of Science in 2010. Source: https://upload.wikimedia.org/wikipedia/commons/thumb/9/97/Professor_Susan_Lindquist_ForMemRS.jpg/220px-Professor_Susan_Lindquist_ForMemRS.jpg*

Preface

NanoBioMedicine is emerging as a potential solution for many medical science problems and will change the face of cancer therapy, diagnostics, imaging, tissue grafting, therapeutics, immunotherapy, and drug delivery in the very near future. Thus, scientists, medical specialists, and individuals with interest in nanotoxicology and nanotechnology will be able to get firsthand knowledge in this field of specialization. The information provided in the book is very concise and crisp and helps to grasp the material with ease.

The book entitled *Nanotoxicology of NanoBioMedicine* will be an essential guide and attract audience from academia, industry, and government from a broad field, including students of medicine, dentistry, veterinary, biology, and toxicology; agricultural scientists; food packaging industrialists; nanotechnologists; as well as nanomaterial scientists and individuals with interest in nanotoxicology and nanotechnology.

The book comprises 10 chapters and presents key concepts of nanotoxicology in biomedicine. Each chapter is concise and crisp that will help the readers to gain an in-depth understanding of the subject of nanotoxicology in nanobiomedicine. Chapter 1 provides extensive coverage of general information, including basic concepts, definitions as relevant to polymer nanoparticles, nonpolymer nanoparticles, nanodevices, hybrid nanomaterials, nanocomposite, nanobiomedical science, and so forth. Chapter 2 deals with historical development from early period to modern nanotechnology including regulations and future trends of nanobiomedicine. Chapter 3 exclusively highlights mechanism of nanomaterial toxicity including pathogenetic pathways activated by oxidative and non-oxidative stress leading to cell death. Chapters 4 and 5 provide overview of factors affecting nanomaterial toxicity, organ- and non-organ-directed nanotoxicity including behavior of NPs in human health and disease, and testing strategies and assessments of potential hazards associated health risk assessments. Chapters 6, 7, and 8 deal with nano-based drug delivery systems; applications of nanotechnology in dentistry, tissue engineering, and regenerative medicine; nanotoxicity of materials used for neurodegenerative disorders, tuberculosis, ophthalmology, respiratory diseases, and various forms of surgical operations in health and disease including their biocompatibility; and toxicity and safety aspects that are important implements to exert control over and monitor the factors that influence drug-laden nanocarriers. Chapter 9 highlights recent advancements regarding tumor targeting, controlled release strategies, and the toxic paradigms of chemotherapeutic drugs and NPs used in cancer therapy. Finally, Chap. 10 summarizes the use of nanomedicine to improve the ability of immunomodulatory molecules to reach diseased tissues, immune cells, or their intracellular compartments, in the context of chronic immune disorders and adverse effects associated with their use. Each chapter is well supported with concise tables, illustrations, diagrams, and important images whenever necessary.

Thus, the book will benefit multiple levels of education including undergraduates, graduate students, and professionals in various fields. This is a cross-functional subject, physicist, material scientist, chemist, biologist, toxicologist, nanotechnologists, biotechnology, medicine including medical practitioners, in various industries such as pharmaceutical, agriculture, chemical, cosmetics and consumer product industries will be the audience for this matter. Of course, this book will also be a good reference for the academic laboratories and students of various disciplines. The currently structured contents of the book would provide a broad knowledge needed to introduce individuals to the field of specialization.

The author welcomes suggestions, constructive criticisms, thoughts, and comments from readers for improving this book. Kindly e-mail to drpkg_brly@yahoo.co.in or to drpkg1943@gmail.com

Bareilly, UP, India   P K Gupta

Disclaimer

The information including illustrations and tables in the book is based on standard textbooks in the area of specialization. However, it is well known that with the advancement of science the standard of care in the practice of biomedicine changes rapidly. Though all the efforts have been made to ensure the accuracy of the information, the possibility of human error still remains. Therefore, neither the author nor the publisher guarantees that the information contained in the book is absolute. Anyone using the clinical relevance information contained in this book has to be, therefore, duly cautious. Neither the author nor the publisher should be responsible for any damage that results from the use of the information contained in any part of this book.

Contents

About the Author

 **P K Gupta** is an internationally known toxicologist with more than 56 years of experience in the field of teaching, research, and research management in the field of toxicology. To his credit, Dr. Gupta has several books and book chapters (John Wiley & Sons; Elsevier, Academic Press; Merck & Co., Mariel Limited; Springer Nature; etc.) as well as scientific research publications (765) published in national and international peer-reviewed journals of repute. He has been a book review editor at Marcel Dekker, New York; an expert member consultant and advisor to WHO, Geneva; a consultant at the United Nations FAO, Rome, and IAEA, Vienna. He has been the founder and past president of the Society of Toxicology of India; founder, director, and member of the nominating committee of the International Union of Toxicology; and founder and editor-in-chief of the peer-reviewed PubMed-indexed scientific journal *Toxicology International*. In addition, Dr. Gupta has also been the biographer for several editions of Who's Who from all over the world, including Marquis Who's Who (USA), IBC (UK), and other leading publications of the world. His name appears in the UP Book of Records and Limca Book of Records for his unique scientific contribution in toxicology. At present, Dr. Gupta has been a former chief of the Division of Pharmacology and Toxicology, Indian Veterinary Research Institute. Presently he is the director of the Toxicology Consulting Group and president of the Academy of Sciences for Animal Welfare.

Chapter 1
Introduction and Historical Background

Abstract Nanomaterials are the particles of organic or inorganic materials and are being used in an ever-increasing number of products and applications. Nano Biomedicine is now being used to help treat diseases and prevent health-related problems. In brief, the prefix "nano" is a Greek word which means "dwarf." The word "nano" means very small or miniature size. This chapter deals with basic information related to nano, an overview of historical development, and various regulations used for the control and use of nanomaterials in drug delivery or controlled release. Finally, the chapter summarizes their future trend in the field of Nano Biomedicine.

Keywords Nano devices · Nanomaterials · Nano medicine · Nano pharmaceutics · Drug delivery · Biosensors · Tissue engineering · History · Current trends · Nano · Regulation · Modern technology · Nano biomedicine

1.1 Introduction

Advancement in the field of nanotechnology and its applications to the fields of medicine and pharmaceuticals have revolutionized the twentieth century. Nanotechnology involves work from the top down, that is, reducing the size of large structures to the smallest structure, for example, photonics applications in nano electronics and nano engineering, top-down or bottom up, which involves changing individual atoms and molecules into nanostructures and more closely resembles chemical-biology. However, it is also inherent that these materials should display different properties such as electrical conductance chemical reactivity, magnetism, optical effects, and physical strength, different from bulk materials due to their small size. This chapter briefly reviews various terminology used and their applications in medicine along with the historical overview of the use of nanoparticles in the ancient and modern era.

© The Author(s), under exclusive license to Springer Nature Switzerland AG 2023
P K Gupta, *Nanotoxicology in Nanobiomedicine*,
https://doi.org/10.1007/978-3-031-24287-8_1

1

1.2   Overview

Nanotechnology is rapidly developing, which leads to the need for safety assessment with regard to both human health and environmental impacts. Materials in the nanoscale can behave differently from larger materials, even if the basic material is the same. Nanoscale materials can have different chemical, physical, electrical, and biological properties. The nanotechnology industry is experiencing challenges in both environmental effects assessment and, therefore, risk assessment. Nanotechnology is the study and control of material which has one or more dimensions in the nanoscale. Nanotechnology is a very multifaceted technology ranging from the extensions of conventional physics to the new approaches and the development of new materials and devices that have at least one dimension in the nanoscale. Nanotechnology also deals with the exploration of whether the material in the nanoscale can be directly measured. Nanomaterials are not a homogenous group of materials but encompass a magnitude of various types and forms of materials having sizes in the range of 1–100 nm. The adverse effects of engineered NMs on living organisms is a rapidly growing discipline aimed at identifying and characterizing nanomaterial toxicity that will serve—in combination with exposure data—the ultimate goal of performing a meaningful risk assessment. Many engineered NMs exhibit unique and desirable catalytic, optical, structural, or electronic properties that make them attractive in diverse technological areas, including multiple manufacturing industries and environmental and medical applications. Therefore, concerns have been raised regarding potential acute and chronic adverse effects due to their physicochemical properties in combination with nanostructures, and the continuing introduction of nanoscale materials into consumer products, such as $TiO2$ in sunscreen creams, antibacterial Ag in textiles, quantum dots in televisions, multiwalled carbon nanotubes (MWCNs) in sports equipment together with increasing numbers of publications reporting toxic responses mostly observed in in vitro studies, has led to increasing concerns and more public awareness about potential adverse health effects. Thus, questions should be raised regarding the actual versus the perceived risk of nanotechnology applications, what is hype, and what is reality. To provide answers it is necessary to understand nanomaterial toxicity in human beings and their environment.

1.3   Nanotoxicology

Nanotoxicology is a new area of study that deals with the toxicological profiles of NMs. Compared with the larger counterparts, the quantum size effects and large surface area to volume ratio brings NMs their unique properties that may or may not be toxic to living things. A combining form with the meaning "very small, minute," it is used in the formation of compound words (nanoplankton); and in the names of units of measure it has a specific sense "one billionth" (10^{-9}): nanomole;

nanosecond. The standard measure of length in science is in meters (m). One nano-meter (1 nm) is equal to 10^{-9} m or 0.000000001 m. A nanometer is 10 times smaller than the width of DNA, and 10 times bigger than the size of an atom.

Comparative terms of prefixes, multiples, and symbols of milli, micro, nano, and pico are summarized as under:

Multiple	Prefix	Symbol
10^{-3}	Milli	m
10^{-6}	Micro	(Greek mu)
10^{-9}	Nano	n
10^{-12}	Pico	p

Nanoparticles (NPs) can reach the blood and may reach other target sites such as the liver, heart, or blood cells. Instead, they may accumulate in biological systems and persist for a long time, which makes such NPs of particular concern. Numerous analytical techniques are available to characterize the toxicological aspects of NPs, but two methods, in particular, are regularly used to grant critical quantitative infor-mation: dynamic light scattering (DLS) and zeta potential (ZP) analysis.

1.4 Nano Biomedicine

Nano medicine refers to the area of science that combines nanotechnology with drugs or diagnostic molecules to improve the ability to target specific cells or tis-sues. Nano medicine ranges from the medical applications of nanomaterials and biological devices. The prefix "nano," derived from the Greek "nanos" signifying "dwarf," is becoming increasingly common in scientific literature. "Nano" is now a popular label for much of modern science, and many "nano-" words have recently appeared in dictionaries, including nanometer, nanoscale, nanoscience, nanotech-nology, nanostructure, nanotube, nanowire, and nanorobot. Many words that are not yet widely recognized are used in several publications. These words include nano-electronics, nanocrystal, nanovalve, nanoantenna, nanocavity, nanoscaffolds, nano-fibers, nanomagnet, nanoporous, nanoarrays, nanolithography, nanopatterning, nanoencapsulation, etc. Although the idea of nanotechnology, producing nanoscale objects and carrying out nanoscale manipulations, has been around for quite some time, the birth of the concept is usually linked to a speech by Richard Feynman at the December 1959 meeting of the American Physical Society where he asked, "What would happen if we could arrange the atoms one by one the way we want them?"

The broad applicability of nanotechnology has led to considerable variation in the terms and definitions used by various scientific communities and regulatory authorities. Despite this, it is generally accepted that NMs have two defining char-acteristics: a size on the nanoscale (typically between 1 nm and 100 nm) and unique size-dependent properties that are not exhibited by the bulk material. For the convenience of the reader, a few of the commonly used terms are defined below:

Nano Biomedicine combines nanotechnology with drugs or diagnostic molecules to improve the ability to target specific cells or tissues. These materials are produced on a nanoscale level and are safe to introduce into the body. Applications and use of nanotechnology in medicine include imaging, tissue engineering, stem cells, *signaling molecules, scaffolds/matrices*, diagnostics (*medical implants, nanoscale sensors,* etc.), and the delivery of drugs that will help medical professionals treat various diseases.

Tissue engineering (TE) is an interdisciplinary field that applies the principles of engineering and the life sciences toward the development of biological substitutes that restore, maintain, or improve tissue function. Bone tissue engineering is concerned with creating implantable bone substitutes for critical skeletal defects that cannot heal on their own. These defects are common clinical scenarios in orthopedics and craniofacial surgery, for the treatment of bone loss due to trauma, infection, and tumor resection.

Stem cells are cells with the potential to develop into many different types of cells in the body. They are unspecialized, so they cannot do specific functions in the body. They have the potential to become specialized cells, such as muscle cells, blood cells, and brain cells. Stem cells originate from two main sources: adult body tissues and embryos. Scientists are also working on ways to develop stem cells from other cells, using genetic "reprogramming" techniques.

Signaling molecules are often called ligands, a general term for molecules that bind specifically to other molecules (such as receptors). The message carried by a ligand is often relayed through a chain of chemical messengers inside the cell.

Scaffolds are materials that have been engineered to cause desirable cellular interactions to contribute to the formation of new functional tissues for medical purposes. Cells are often "seeded" into these structures capable of supporting three-dimensional tissue formation. A bone scaffold is a 3D matrix that allows and stimulates the attachment and proliferation of osteoinducible cells on its surfaces. Various synthetic and natural, and biodegradable and non-biodegradable materials have been used in the fabrication of bone scaffolds through different methods.

Diagnostics such as medical implants, and knee and hip replacements have improved the lives of millions, but a common problem with these implants is the risk of post-surgery inflammation and infection. In many cases, symptoms from an infection are detected so late that treatment is less effective, or the implant will need to be replaced altogether.

Nanoscale sensors embedded directly into the implant or surrounding area could detect infection much sooner. As targeted drug delivery becomes more feasible, it could be possible to administer treatment to an infected area at the first sign of infection. Before long, gathering data from within the body and administering treatments in real-time could move from science fiction to the real world.

Targeted drug delivery, sometimes called smart drug delivery, is a method of delivering medication to a patient in a manner that increases the concentration of the medication in some parts of the body relative to others. This means of delivery is largely founded on nano medicine, which plans to employ NP-mediated drug delivery in order to combat the downfalls of conventional drug delivery. These nanoparticles would be loaded with drugs and targeted to specific parts of the body where there is solely diseased tissue, thereby avoiding interaction with healthy tissue. The goal of a targeted drug delivery system is to prolong, localize, target, and have a protected drug interaction with the diseased tissue.

Nano-drug delivery systems (NDDSs) are a class of nanomaterials that have abilities to increase the stability and water solubility of drugs, prolong the cycle time, increase the uptake rate of target cells or tissues, and reduce enzyme degradation, thereby improving the safety and effectiveness of drugs.

A *nanocarrier* is nanomaterial being used as a transport module for another substance, such as a drug. Commonly used nanocarriers include micelles, polymers, carbon-based materials, liposomes, and other substances. Nanocarriers are currently being studied for their use in drug delivery and their unique characteristics demonstrate potential use in chemotherapy.

Bioactive materials are nontoxic, biologically active materials that induce the formation of a direct chemical bond between the implant and host tissue by eliciting a biological response at the interface.

Bioglass: The primary application of bioglass is the repair of bone injuries or defects too large to be regenerated by the natural process. The first successful surgical use of bioglass 45S5 was in replacement of ossicles in the middle ear, as a treatment for conductive hearing loss.

Antibody-drug conjugates or ADCs are a class of biopharmaceutical drugs designed as a targeted therapy for treating cancer. Unlike chemotherapy, ADCs are intended to target and kill tumor cells while sparing healthy cells. ADCs are complex molecules composed of an antibody linked to a biologically active cytotoxic (anticancer) payload or drug. Figure 1.1 is a simplified illustration of the mechanisms by which an antibody-drug conjugate (ADC) may be internalized by a hepatocellular cancer (HCC) cell by targeting the cell surface membrane-bound Glypican-3 (GPC3), a heparan sulfate proteoglycan protein. GPC3 internalization is triggered by its binding to the hedgehog (Hh) polypeptide ligand by a mechanism involving low-density-lipoprotein receptor-related protein-1 (LRP1). An ADC bound to GPC3 can be degraded together with GPC3 following its internalization, which triggers the release of the active drug. Antibody-drug conjugates are examples of bioconjugates and immunoconjugates.

ADCs combine the targeting capabilities of monoclonal antibodies with the cancer-killing ability of cytotoxic drugs. They can be designed to discriminate between healthy and diseased tissue.

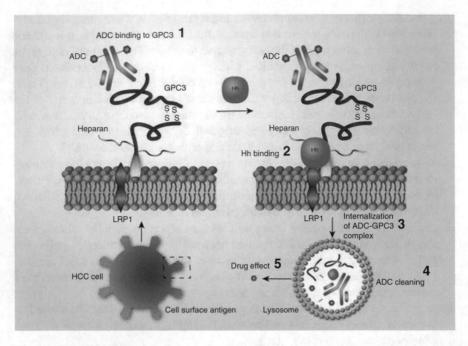

Fig. 1.1 Simplified illustration of the mechanisms of biopharmaceutical drugs designed as a targeted therapy for treating cancer
https://www.mdpi.com/molecules/molecules-25-02861/article_deploy/html/images/molecules-25-02861-g003-550.jpg

1.5 Nanoelectromechanical Systems

The acronym NEMS stands for nanoelectromechanical systems. NEMS are a class of devices integrating electrical and mechanical functionality on the nanoscale. NEMS form the next logical miniaturization step from so-called microelectromechanical systems, or MEMS devices. NEMS typically integrate transistor-like nanoelectronics with mechanical actuators, pumps, or motors, and may thereby form physical, biological, and chemical sensors.

1.6 Microelectromechanical Systems

The acronym MEMS stands for microelectromechanical systems, but MEMS generally refers to microscale devices or miniature embedded systems involving one or more micromachined component that enables higher-level functionality.

1.7   Nanotechnology: Structures

This is the branch of technology that deals with making structures that are less than 100 nanometers long. Scientists often build these structures using individual molecules of substances. This technology involves manipulation and control of matter on the nanoscale dimension by using scientific knowledge of various industrial and biomedical applications. NMs are materials with a microstructure the characteristic length scale of which is on the order of a few (typically 1–100) nanometers. These materials are of different types such as polymer NPs; *carbon nanotubes* (CNTs); metal NPs; nanoalloys or their compounds; quantum dots; silica particles; nanodevices; hybrid and composites (nanocomposite, hybrid nanocomposite, ceramic matrix composites, polymer-matrix nanocomposites); magnetic nanocomposites; heat-resistant nanocomposites; and so on.

Polymer NP that may be *organic NPs* of two or more nanoscale dimensions, with a size in the range of 1–100 nm. The chemical composition and the shape of a NP also influence its specific properties. The main groups of *organic* nano carriers are liposomes, micelles, protein/peptide based and dendrimers. The terms "nanorod" or "nanoplate" are employed, instead of nanoparticle, when the longest and the shortest axes lengths of a nano-object are different.

CNTs are *tubes* made of *carbon* with diameters typically measured in nanometers. *CNTs* often refer to single-wall *carbon nanotubes* (SWCNTs) with diameters in the range of a nanometer. CNTs can act as antennas for radios and other electromagnetic devices. Conductive CNTs are used in brushes for commercial electric motors.

Metal NPs are submicron scale entities made of pure metals (e.g., gold, platinum, silver, titanium, zinc, cerium, iron, and thallium) or their compounds (e.g., oxides, hydroxides, sulfides, phosphates, fluorides, and chlorides). Apart from catalysis, potential applications of metal NPs are well known in other fields such as pigments, electronic and magnetic materials, and drug delivery.

Quantum dots are tiny particles or nanocrystals of a semiconducting material with diameters in the range of 2–10 nm (10–50 atoms), having optical and electronic properties that differ from larger particles due to quantum mechanics. When the quantum dots are illuminated by UV light, an electron in the quantum dot can be excited to a state of higher energy. In the case of a semiconducting quantum dot, this process corresponds to the transition of an electron from the valence band to the conductance band. The excited electron can drop back into the valence band releasing its energy by the emission of light.

Silica, or silicon dioxide, is the same material used to make glass. In nature, silica makes up quartz and the sand you walk on at the beach. Unlike metals such as gold and iron, silica is a poor conductor of both electrons and heat. Despite these limitations, silica (silicon oxide) NPs form the framework of silica aerogels. These properties make nano aerogels one of the best thermal insulators known to man.

Nanodevices are NPs that are created for the purpose of interacting with cells and tissues and carrying out very specific tasks. The most famous nanodevices are the imaging tools. Oral pills can be taken that contain miniature cameras.

Nanocomposites are heterogeneous/hybrid materials that are produced by the mixtures of polymers with inorganic solids (clays to oxides) at the nanometric scale. Most popularly, nanocomposites are prepared by the process within in situ growth and polymerization of biopolymer and inorganic matrix. The idea behind nanocomposite is to use building blocks with dimensions in nanometer range to design and create new materials with unprecedented flexibility and improvement in their physical properties. Nanocomposites are fire-resistant or flame retardant and have accelerated biodegradability. A range of polymeric nanocomposites are used for biomedical applications, such as tissue engineering, drug delivery, cellular therapies, and for nanoelectronic biosensors, and even possible future applications of molecular nanotechnology such as biological machines.

Hybrid nanocomposite materials are unique chemical conjugates of organic and/ or inorganic materials. That is, these are mixtures of two or more inorganic components, two or more organic components, or at least one of both types of components. For example, structural nanocomponents such as a liposome, a micelle, mesoporous silica, a polymer, or a virus can mainly carry a drug cargo, while structural nanocomponents such as a gold nanoparticle or a carbon nanotube enable photoablation therapy.

Ceramic matrix composites (CMCs) consist of ceramic fibers embedded in a ceramic matrix. The matrix and fibers can consist of any ceramic material, including carbon and carbon fibers. The ceramic occupying most of the volume is often from the group of oxides, such as nitrides, borides, and silicides, whereas the second component is often a metal. Ideally both components are finely dispersed in each other in order to elicit particular optical, electrical, and magnetic properties as well as tribological, corrosion-resistance, and other protective properties.

Metal matrix nanocomposites can also be defined as reinforced metal matrix composites. This type of composite can be classified as continuous and noncontinuous reinforced materials. One of the more important nanocomposites is carbon nanotube metal matrix composites, which is an emerging new material that is being developed to take advantage of the high tensile strength and electrical conductivity of carbon nanotube materials.

Appropriately adding nanoparticulates to a polymer matrix can enhance its performance, often dramatically, by simply capitalizing on the nature and properties of the nanoscale filler (these materials are better described by the term "nanofilled polymer composites"). This strategy is particularly effective in yielding high-performance composites, when uniform dispersion of the filler is achieved and the properties of the nanoscale filler are substantially different or better than those of the matrix.

Magnetic nanocomposites that can respond to an external stimulus are of increased interest due to the fact that, because of the large amount of interaction between the phase interfaces, the stimulus response can have a larger effect on the composite as a whole. The external stimulus can take many forms, such as a magnetic, electrical, or mechanical field. Specifically, magnetic nanocomposites are useful for use in these applications due to the nature of magnetic material's ability to respond both to electrical and magnetic stimuli. The penetration depth of a magnetic field is also high, leading to an increased area that the nanocomposite is affected by and therefore an increased response. In order to respond to a magnetic field, a matrix can be easily loaded with nanoparticles or nanorods.

In recent years, nanocomposites have been designed to withstand high temperatures by the addition of carbon dots (CDs) in the polymer matrix. Such nanocomposites can be utilized in environments wherein high temperature resistance is a prime criterion.

1.8 Historical Development

1.8.1 Early Period

The practice of nanotechnology can be traced to ancient times when human characteristics of curiosity, wonder, and ingenuity are as old as mankind. People around the world have been harnessing their curiosity into inquiry and the process of scientific methodology. Even in 5000 BC, clay was used to bleach wools and clothes in Cyprus. Humans already exploited the reinforcement of ceramic matrixes by including natural asbestos nanofibers more than 4500 years ago. The Ancient Egyptians were also using NMs more than 4000 years ago based on a synthetic chemical process to synthesize ≈ 5 nm diameter PbS NPs for hair dye. Similarly, "Egyptian blue" was the first synthetic pigment which was prepared and used by Egyptians using a sintered mixture nanometer-sized glass and quartz around the third century BC. Egyptian blue represents a multifaceted mixture of $CaCuSi_4O_{10}$ and SiO_2 (both glass and quartz).

Even long before the start of "nanoera," people were coming across various nanosized objects and the related nanolevel processes and using them in practice. However, intuitive nanotechnology antiquities developed spontaneously, without due understanding of the nature of these objects and processes. For example, the fact that small particles of various substances possessed properties different to those of the same substances with larger particle size was known for a long time, but the reason for this was not clear. Thus, people were engaged in nanotechnology subconsciously, without guessing that they were dealing with the nanoworld phenomena. In many instances secrets of ancient nanoproduction simply passed from generation to generation, without getting into the reasons why the received materials and products derived from them acquired their unique properties. Since decades (1000 BC) people knew and used natural fabrics: flax, cotton, wool, and silk. They were able to cultivate them and process into products. What makes these fabrics special is the fact that they have a developed network of pores with the size of 1–20 nm, that is, they are typical nanoporous materials. Due to their nanoporous structure natural fabrics possess high utilitarian properties: They absorb sweat well, quickly swell, and dry. Since ancient times people mastered the ways of making bread, wine, beer, cheese, and other foodstuffs, where the fermentation processes on nanolevel are critical. In Ancient Egypt, it was rather common to dye hair black. For a long time it was believed that the Egyptians used mainly natural vegetative dyes—henna and black hair dye. However, recent research into hair samples from ancient Egyptian burial sites, conducted by Ph. Walter, showed that hair was dyed in black with paste from lime, lead oxide, and a small amount of water. In the course of the dyeing process nanoparticles of galenite (lead sulfide) were formed. Natural black hair color is provided with a pigment called melanin, which in the form of inclusions is spread in hair keratin. The Egyptians were able to make the dyeing paste react with sulfur, which is part of keratin, and receive galenite particles a few nanometers in size, which provided even and steady dyeing. The British museum boasts Licurg's bowl as part of its heritage—an outstanding product of glass makers of Ancient Rome. This bowl, on which Licurg, the tsar of Edons, is depicted, possesses unusual optical properties: It changes color with change of location (inside or outside) of the light source. In natural light the bowl is green, and if illuminated from within, it turns red.

1.8.2   Premodern Period

Our knowledge of nanotechnology during the fourth to fifteenth centuries, the premodern times, is derived from a long list of discoveries from early examples of nanostructured materials which were based on craftsmen's empirical understanding and manipulation of materials. Use of high heat was one common step in their processes to produce these materials with novel properties. The examples include:

The Lycurgus Cups are a fourth-century Roman glass cup, made of a dichroic glass that displays different colors: red when a light passes from behind and green when a light passes from the front. Recent studies found that the Lycurgus Cups contain Ag–Au alloy NPs, with a ratio of 7:3 in addition to about 10% Cu. Subsequently, red and yellow colored stained glass found in medieval period churches was produced by incorporating colloidal Au and Ag NPs, respectively (Fig. 1.2).

These decorations showed amazing optical properties due to the existence of distinct Ag and/or Cu NPs isolated within the outermost glaze layers. These decorations are an example of metal NPs that display iridescent bright green and blue colors under particular reflection conditions. TEM analysis of these ceramics revealed a double layer of Ag NPs (5–10 nm) in the outer layer and larger ones (5–20 nm) in the inner layer. The distance was observed to be constant at about 430 nm in between two layers, giving rise to interference effects. The scattered light from the second layer leads to the phase shift due to the scattering of light by the first layer. This incoming light wavelength-dependent phase shift leads to a different wavelength while scattering.

Fig. 1.2 The Lycurgus
Cup at the British
Museum, lit from the
outside (*left*) and from the
inside (*right*)
https://www.nano.gov/sites/
default/files/lycurguscup1a.jpg

1.8.3   Middle Ages

Our knowledge of nanoparticles during the ninth to seventeenth centuries, the Middle Ages, is derived from a list of articles. For example, glowing, glittering "luster" ceramic glazes used in the Islamic world, and later in Europe, contained silver or copper or other metallic nanoparticles (Fig. 1.3), and vibrant stained glass windows in European cathedrals owed their rich colors to nanoparticles of gold chloride and other metal oxides and chlorides; gold nanoparticles also acted as photocatalytic air purifiers.

Likewise, during the thirteenth to eighteenth centuries, "Damascus" saber blades contained carbon nanotubes and cementite nanowires—an ultrahigh-carbon steel formulation that gave them strength, resilience, the ability to hold a keen edge, and a visible moiré pattern in the steel that give the blades their name. Damascus steel was the forged steel of the blades of swords smithed in the Near East from ingots of Wootz steel either imported from Southern India or made in production centers in Sri Lanka, or Khorasan, Iran. These swords are characterized by distinctive patterns of banding and mottling reminiscent of flowing water, sometimes in a "ladder" or "rose" pattern. Such blades were reputed to be tough, resistant to shattering, and capable of being honed to a sharp, resilient edge. A sword maker of Damascus, Syria, is a well-known example of nanoparticle technology of Middle Ages.

Actually, Damasqui was a sword maker, one in a long line of smiths who forged the legendary weapons known as Damascus sabers. They were strong yet flexible and supremely sharp, which European warriors first discovered, much to their misfortune, at the hands of Muslims during the Crusades.

The recipe for making Damascus steel was lost at the end of the eighteenth century, so no one knew the reasons for its remarkable qualities. But an analysis by

Fig. 1.3 Polychrome lusterware bowl, ninth C, Iraq, British Museum
https://www.nano.gov/sites/default/files/iraq_polychromelustrewarebowl-archaeology.com-s.jpg

twenty-first-century researchers in Germany provides a clue: Damascus sabers, they report in *Nature*, contain carbon nanotubes.

Using a transmission electron microscope, Peter Paufler of the Technical University of Dresden and colleagues looked at a very thin sample of steel from a saber made by Assad Ullah, who worked in what is now Iran. What they saw seemed for all the world like carbon nanotubes, cylindrical arrangements of carbon atoms first discovered in 1991 and now made in laboratories all over the world. Further analysis confirmed that that was what they were.

1.8.4 Modern Nanotechnology

In 1857, Michael Faraday discovered colloidal "ruby" gold, demonstrating that nanostructured gold under certain lighting conditions produces different-colored solutions, which is the first scientific description to report NP preparation and initiated the history of NMs in the scientific arena. He also revealed that the optical characteristics of Au colloids are dissimilar compared to their respective bulk counterpart. This was probably one of the earlier reports where quantum size effects were observed and described.

In 1908, Mie described the scattering of an electromagnetic plane wave by a homogeneous sphere (Mie scattering). The solution takes the form of an infinite series of spherical multipole partial waves. It is named after Gustav Mie (Fig. 1.4). In this sunrise image, the blue sky, yellow Cirrus clouds, and orange Altocumulus

Fig. 1.4 Mie scattering, artistic view: In this sunrise image, the blue sky, yellow Cirrus clouds and orange Altocumulus clouds result from both Rayleigh and Mie scattering. Rayleigh scattered produces blue sky and the color the clouds receives. Mei scattering is responsible for the color we see. Even with Rayleigh scattering taking place in the atmosphere, over one-half of the sun's 'white' light continues through the atmosphere reaching the earth's surface
https://www.weather.gov/images/jetstream/clouds/sunrise_colors.jpg

clouds result from both Rayleigh and Mie scattering. Rayleigh scattered produces blue sky and the color the clouds receive. Mei scattering is responsible for the color we see. Even with Rayleigh scattering taking place in the atmosphere, over one-half of the sun's "white" light continues through the atmosphere reaching the earth's surface.

Mie scattering (sometimes referred to as a non-molecular scattering or aerosol particle scattering) takes place in the lower 4500 m (15,000 ft) of the atmosphere, where there may be many essentially spherical particles present with diameters approximately equal to the size of the wavelength of the incident ray. Mie scattering theory has no upper size limitation and converges to the limit of geometric optics for large particles. Mie scattering, artistic view:

In the 1940s, SiO_2 NPs were manufactured as substitutes to carbon black for rubber reinforcement. Later, a similar technique was used to produce the famous Satsuma glass in Japan. The absorption properties of Cu NPs are helpful in brightening the Satsuma glass with ruby color. Furthermore, clay minerals with a thickness of a few nanometers are the best examples of natural NM usage for various vessels since antiquity. Subsequently, the analysis of fragments of the bowl, carried out in the laboratories of General Electric in 1959 for the first time, showed that the bowl consists of usual soda-lime-quartz glass and has about 1% of gold and silver, and also 0.5% of manganese as components. The researchers then assumed that the unusual color and disseminating effect of glass is provided by colloidal gold. Later, when research techniques became more advanced, scientists discovered particles of gold and silver from 50 to 100 nm in size using an electronic microscope.

Today manufactured NMs can significantly improve the characteristics of bulk materials, in terms of strength, conductivity, durability, and lightness, and they can provide useful properties (e.g., self-healing, self-cleaning, anti-freezing, and anti-bacterial) and can function as reinforcing materials for construction or sensing components for safety. Notwithstanding the other possible benefits, simply taking advantage of the beneficial size and shape effects to improve the appearance of materials is still a major application of NPs. Moreover, the commercial use of NMs is often limited to the bulk use of passive NMs embedded in an inert (polymer or cement) matrix, forming a nanocomposite. In 2003, Samsung introduced an anti-bacterial technology with the trade name Silver Nano™ in their washing machines, air conditioners, refrigerators, air purifiers, and vacuum cleaners, which use ionic Ag NPs. NPs and nanostructured materials are extensively used in auto production: as fillers in tires to improve adhesion to the road, fillers in the car body to improve the stiffness, and as transparent layers used for heated, mist, and ice-free window panes. By the end of 2003, Mercedes-Benz brought a NP-based clear coat into series production for both metallic and nonmetallic paint finishes. The coating increases the scratch resistance and enhances the gloss. Liquid magnets, so-called ferrofluids, are ultrastable suspensions of small magnetic NPs with superparamagnetic properties. Upon applying a magnetic field, the liquid will macroscopically magnetize, which leads to the alignment of NPs along the magnetic field direction. In 2005, Abraxane™, which is a human serum albumin NP material containing paclitaxel, was manufactured, commercialized, and released in the pharmaceutical market. In 2006 P. Paufler carried out the research of saber fragments made from the Damask steel using an electronic microscope. The results showed that the steel had a nanofibrous structure. It is supposed that such a structure was received after a special thermomechanical processing of steel, which was made from the ore of a special structure.

In the summer of 2012, Logitech brought an external iPad keyboard powered by light on the market, representing the first major commercial use of dye-sensitized solar cells. In 2014, there were about 1814 nanotechnology-based consumer products that are commercially available in over 20 countries.

Periodical development in nanotechnology starting from 1857 and the various stages of it have been summarized as under:

1857: Michael Faraday discovered colloidal "ruby" gold, demonstrating that nano-structured gold under certain lighting conditions produces different-colored solutions.

1936: Erwin Müller, working at Siemens Research Laboratory, invented the field emission microscope, allowing near-atomic-resolution images of materials.

1947: John Bardeen, William Shockley, and Walter Brattain at Bell Labs discovered the semiconductor transistor and greatly expanded scientific knowledge of semiconductor interfaces, laying the foundation for electronic devices and the Information Age.

1950: Victor La Mer and Robert Dinegar developed the theory and a process for growing monodisperse colloidal materials. The controlled ability to fabricate colloids enables myriad industrial uses such as specialized papers, paints, and thin films, even dialysis treatments.

1951: Erwin Müller pioneered the field ion microscope, a means to image the arrangement of atoms at the surface of a sharp metal tip; he first imaged tungsten atoms.

1956: Arthur von Hippel at MIT introduced many concepts of—and coined the term—"molecular engineering" as applied to dielectrics, ferroelectrics, and piezoelectrics.

1958: Jack Kilby of Texas Instruments originated the concept of, designed, and built the first integrated circuit, for which he received the Nobel Prize in 2000.

1959: Richard Feynman of the California Institute of Technology gave what is considered to be the first lecture on technology and engineering at the atomic scale, "There's Plenty of Room at the Bottom," at an American Physical Society meeting at Caltech (image at right).

1965: Intel co-founder Gordon Moore described in *Electronics* magazine several trends he foresaw in the field of electronics. One trend is now known as "Moore's Law."

1974: Tokyo Science University Professor Norio Taniguchi coined the term "nanotechnology" to describe precision machining of materials to within atomic-scale dimensional tolerances.

1981: Gerd Binnig and Heinrich Rohrer at IBM's Zurich lab invented the scanning tunneling microscope, allowing scientists to "see" (create direct spatial images of) individual atoms for the first time. Binnig and Rohrer won the Nobel Prize for this discovery in 1986.

1985: Rice University researchers Harold Kroto, Sean O'Brien, Robert Curl, and Richard Smalley discovered the Buckminsterfullerene (C60), more commonly known as the buckyball (Fig. 1.5), which is a molecule resembling a soccer ball in shape and composed entirely of carbon, as are graphite and diamond. The team was awarded the 1996 Nobel Prize in Chemistry for their roles in this discovery and that of the fullerene class of molecules more generally.

1985: Bell Labs' Louis Brus discovered colloidal semiconductor nanocrystals (quantum dots), for which he shared the 2008 Kavli Prize in Nanotechnology.

1986: Gerd Binnig, Calvin Quate, and Christoph Gerber invented the atomic force microscope, which has the capability to view, measure, and manipulate materials down to fractions of a nanometer in size, including measurement of various forces intrinsic to nanomaterials.

1986: First book on nanotechnology *Engines of Creation* published by K. Eric Drexler, Atomic Force Microscope.

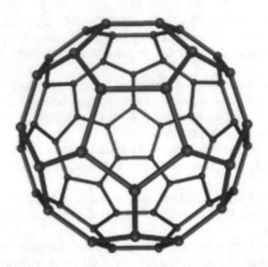

Fig. 1.5 Buckminsterfullerene (C60), more commonly known as the buckyball

1989: Don Eigler and Erhard Schweizer at IBM's Almaden Research Center manipulated 35 individual xenon atoms to spell out the IBM logo. This demonstration of the ability to precisely manipulate atoms ushered in the applied use of nanotechnology.

1990s: Early nanotechnology companies began to operate, for example, Nanophase Technologies in 1989, Helix Energy Solutions Group in 1990, Zyvex in 1997, Nano-Tex in 1998.

1991: Sumio Iijima of NEC Corporation of Japan is credited with discovering the carbon nanotube (CNT), although there were early observations of tubular carbon structures by others as well. Iijima shared the Kavli Prize in Nanoscience in 2008 for this and other advances in the field. CNTs, like buckyballs, are entirely composed of carbon, but in a tubular shape. They exhibit extraordinary properties in terms of strength, electrical and thermal conductivity, among others.

1992: C.T. Kresge and colleagues at Mobil Oil discovered the nanostructured catalytic materials MCM-41 and MCM-48 {Mobil Composition of Matter (MCM) is the initial name given for a series of mesoporous materials that were first synthesized by Mobil's researchers in 1992}, now used heavily in refining crude oil as well as for drug delivery, water treatment, and other varied applications.

1993: Moungi Bawendi of MIT invented a method for controlled synthesis of nanocrystals (quantum dots), paving the way for applications ranging from computing to biology to high-efficiency photovoltaics and lighting.

1998: The Interagency Working Group on Nanotechnology (IWGN) was formed under the National Science and Technology Council to investigate the state of the art in nanoscale science and technology and to forecast possible future developments. The IWGN's study and report, Nanotechnology Research Directions: Vision for the Next Decade (1999), defined the vision.

1999: First nano medicine book by R. Freitas *Nano medicine* was published.

1999: Cornell University researchers Wilson Ho and Hyojune Lee probed secrets of chemical bonding by assembling a molecule [iron carbonyl Fe(CO)2] from constituent components [iron (Fe) and carbon monoxide (CO)] with a scanning tunneling microscope.

1999: Chad Mirkin at Northwestern University invented dip-pen nanolithography® (DPN®), leading to manufacturable, reproducible "writing" of electronic circuits as well as patterning of biomaterials for cell biology research, nanoencryption, and other applications.

1999–early 2000s: Consumer products making use of nanotechnology began appearing in the marketplace.

2000: President Clinton launched the National Nanotechnology Initiative (NNI) to coordinate Federal R&D efforts and promote US competitiveness in nanotechnology.

2001: For developing theory of nanometer-scale electronic devices and for synthesis and characterization of carbon nanotubes and nano wires, Feynman Prize in Nanotechnology was awarded.

2002: Feynman Prize in Nanotechnology was awarded for using DNA to enable the self-assembly of new structures and for advancing our ability to model molecular machine systems.

2003: Feynman Prize in Nanotechnology was awarded for modeling the molecular and electronic structures of new materials and for integrating single molecule biological motors with nanoscale silicon devices.

2003: Congress enacted the twenty-first century Nanotechnology Research and Development Act (P.L. 108 153). The act provided a statutory foundation for the NNI, established programs, assigned agency responsibilities, authorized funding levels, and promoted research to address key issues.

2004: The European Commission adopted the Communication "Towards a European Strategy for Nanotechnology."

2004: First policy conference on advanced nanotech was held. First center for nano mechanical systems was established, and Feynman Prize in Nanotechnology was awarded for designing stable protein structures and for constructing a novel enzyme with an altered function.

2004: SUNY Albany launched the first college-level education program in nanotechnology in the United States, the College of Nanoscale Science and Engineering.

2005: Erik Winfree and Paul Rothemund from the California Institute of Technology developed theories for DNA-based computation and "algorithmic self-assembly" in which computations are embedded in the process of nanocrystal growth.

2006: James Tour and colleagues at Rice University built a nanoscale car made of oligo (phenylene ethynylene) with alkynyl axles and four spherical C60 fullerene (buckyball) wheels.

2007: Angela Belcher and colleagues at MIT built a lithium-ion battery with a common type of virus that is non-harmful to humans, using a low-cost and environmentally benign process.

2008: The first official National Nanotechnology Initiative (NNI) Strategy for Nanotechnology-Related Environmental, Health, and Safety (EHS) Research was published.

2009–2010: Nadrian Seeman and colleagues at New York University created several DNA-like robotic nanoscale assembly devices. One is a process for creating 3D DNA structures using synthetic sequences of DNA crystals that can be programmed to self-assemble using "sticky ends" and placement in a set order and orientation.

2011: The Nanoscale Science, Engineering, and Technology (*NSET*) subcommittee updated both the NNI Strategic Plan {The National Nanotechnology Initiative (NNI) is a US Government research and development (R&D) initiative involving the nanotechnology-related activities of 20 departments and independent agencies. The NNI consists of the individual and cooperative nanotechnology-related activities of federal agencies with a range of research and regulatory roles and responsibilities.} and the NNI Environmental, Health, and Safety Research Strategy.

2012: The NNI launched two more Nanotechnology Signature Initiatives (NSIs)— Nanosensors and the Nanotechnology Knowledge Infrastructure (NKI)—bringing the total to five NSIs.

2013: The NNI started Strategic Planning, starting with the Stakeholder Workshop.

– Stanford researchers developed the first carbon nanotube computer.

2014: The NNI released the updated 2014 Strategic Plan.

– The NNI released the *2014 Progress Review on the Coordinated Implementation of the NNI 2011 Environmental, Health, and Safety Research Strategy.*

The period (2011–2020) has mainly focused on nanoscale science and engineering integration, science-based design of fundamentally new products, and general-purpose and mass use of nanotechnology. The focus of R&D and applications is expected to shift toward more complex nanosystems, new areas of relevance, and fundamentally new products. This phase is expected to be dominated by an R&D ecosystem driven by socio-economic considerations; it might be called "Nano 2."

1.9   Regulations

Nanomaterials possess characteristics such as high chemical bioactivity and reactivity, cellular as well as tissue and organ penetration ability, and greater bioavailability. These unique properties of NMs make them superior in biomedical applications. These merits are also avenues for potential toxicity. Thus, regulations via legislation, laws, and rules have been implemented by several government organizations to minimize or avoid risks associated with NMs. However, there is no specific international regulation, no internationally agreed upon protocols or legal definitions for production, handling or labeling, testing toxicity and evaluating the environmental impact of NPs.

Medical standards related to ethics, environmental safety, and medical governance have been modified to cover the introduction of NMs into the biomedical field. Currently, the US and the European Union (EU) have strong regulatory bodies and guideline legislation to control the potential risks of NMs. The European Commission has developed several pieces of EU legislation and technical guidance, with specific references to NMs. This legislation has been employed inside EU countries to ensure conformity across legislative areas and to guarantee that a NM in one sector will also be treated as such when it is used in another sector. According to the European Commission the term "nanomaterial" means "a natural, incidental or manufactured material containing particles, in an unbound state or as an aggregate or as an agglomerate, and where for 50% or more of the particles in the number size distribution, one or more external dimensions is in the size range of 1 nm to 100 nm." As the specifications of the materials and products meet the substance definitions of the European chemical agency (REACH) and the European Classification and Labelling of Chemicals (CLP), the provisions in these regulations apply. In addition, the EU has formed the Scientific Committee on Emerging and Newly Identified Health Risks (SCENIHR), to estimate risks associated with NMs. In 2013, EU cosmetics regulation 1223/2009 was replaced by Directive 76/768/EEC. The regulation defines the term "nanomaterial" as "an insoluble or biopersistent and intentionally manufactured material with one or more external dimensions, or an internal structure in the range of 1 to 100 nm which includes man-made fullerene, single-walled carbon nanotubes, and graphene flakes." Cosmetics face regulations and moderations from USFDA's Federal Food, Drug, and Cosmetic Act (FFDCA), Personal Care Products Council (PCPC), Voluntary Cosmetic Registration Progam (VCRP), EU cosmetics product notification portal (CPNP), REACH, Scientific Committee on Consumer Safety (SCCS), and International Cooperation on Cosmetic Regulation (ICCR). These regulations from the US and EU, as well as other countries such as Japan and Canada, reveal that nanotoxicity via cosmetics are of major concern for both scientific policymakers and industries producing consumer products.

In the US, regulatory agencies such as the Food and Drug Administration (FDA), the United States Environmental Protection Agency (USEPA), and the Institute for Food and Agricultural Standards (IFAS) have initiated protocols to deal with the possible risks of NMs and nanoproducts. Since 2006, the FDA has been working on identifying sources of NMs, estimating the environmental impact of NMs and their risks on people, animals, and plants and how these risks could be avoided or mitigated.

The European Medicines Agency (EMEA) and United States Food and Drug Administration (USFDA) help in regulating the medical usage of hazardous NMs. Apart from this, a book entitled *Principles for the Oversight of Nanotechnologies and Nanomaterials* was published by a coalition of US domestic and international advocacy groups and was endorsed by 70 groups on six continents. This article demands for a strong and comprehensive oversight of products generated from

NMs. This encompasses a precautionary foundation for specific nanomaterial regulations, health and safety of the public and workers, transparency, public participation, environmental protection, as well as the inclusion of broader impacts and manufacturer liability. Similarly, the Nanomaterials Policy Recommendations report covers ways to avoid or reduce the risk of NMs in food-related industries. This report also advises companies to adopt a detailed public policy for NMs usage, publish safety analyses of NMs, issue supplier standards, label NPs below 500 nm, and adopt a hazard control approach to prevent exposure to NPs. Organic suppliers including the UK Soil Association, the Biological Farmers of Australia, and the Canada General Standards Board have already banned the use of engineered NPs in food. Researchers and manufacturers should be educated on the regulatory laws and legislations prior to NMnanomaterial production to avoid these types of bans against NMs. It is currently agreed that NMs are not intrinsically hazardous per se and many of them seem to be nontoxic, while others have beneficial health effects. However, the risk assessment in the future will determine whether the NMs and their products are hazardous or any further actions are needed.

1.10   Future Trends

In 1999/2000, a convergence was reached in defining the nanoscale world because typical phenomena in material nanostructures were better measured and understood with a new set of tools, and new nanostructures had been identified at the foundations of biological systems, nanomanufacturing, and communications. The new challenge for the next decade is building systems from the nanoscale that will require the combined use of nanoscale laws, biological principles, information technology, and system integration. Because of their unique properties, nanomaterials pose new and yet unknown risks to human health and the environment. These microscopic materials that are increasingly being integrated into food such as baby formula and common candies, as well as food storage containers and cookware, have unprecedented mobility. This allows them to penetrate human skin and when ingested, reach sensitive places like bone marrow, lymph nodes, the heart, and the brain. Despite these novel and dangerous properties, nanomaterials are subject to the regulatory system governing larger materials of the same substance. No additional regulations have been put in place to account for the differing behavior of nanomaterials. Despite the absence of a sufficient regulatory framework or proper method of evaluation, substances including nanosilver, nanozinc, and nanosilica are increasingly being found in a wide variety of consumer goods. In the future, one may expect divergent trends as a function of system architectures. Several possible divergent trends are system architectures based on guided molecular and macromolecular assembling, robotics, biomimetics, and evolutionary approaches.

Around the world, researchers are increasingly thinking smaller to solve some of the biggest problems in medicine. Though most biological processes happen at the nano level, it wasn't until recently that new technological advancements helped in opening up the possibility of nano medicine to healthcare researchers and professionals.

1.10.1   Small Systems, Big Applications

While smart pill technology is not a new idea—a "pill cam" was cleared by the FDA in 2001—researchers are coming up with innovative new applications for the concept. For example, MIT researchers designed an ingestible sensor pill that can be wirelessly controlled. The pill would be a "closed-loop monitoring and treatment" solution, adjusting the dosage of a particular drug based on data gathered within the body (e.g., gastrointestinal system). An example of this technology in action is the recent FDA-approved smart pill that records when medication was taken. The product, which is approved for people living with schizophrenia and bipolar disorder, allows patients to track their own medication history through a smartphone, or to authorize physicians and caregivers to access that information online.

1.10.2   Beating the Big C

Nearly 40% of humans will be diagnosed with cancer at some point in their lifetime, so any breakthrough in cancer treatment will have a widespread impact on society. On the key issues with conventional chemotherapy and radiation treatments is that the body's healthy cells can become collateral damage during the process. For this reason, researchers around the world are working on using nano particles to specifically target cancer cells. Oncology-related drugs have the highest forecasted worldwide prescription drug sales, and targeting will be a key element in the effectiveness of these powerful new drugs.

1.10.3   Scientific Trends in Medicine

Improving the ability of nanotechnologies to target specific cells or tissues is of great interest to companies producing nano medicines. This area of research involves attaching nanoparticles onto drugs or liposomes to increase specific localization. Since different cell types have unique properties, nanotechnology can be used to "recognize" cells of interest. This allows associated drugs and therapeutics to reach diseased tissue while avoiding healthy cells. While this is a promising area of research, very few nano medicines exist that successfully utilize nanotechnology in this manner. This is due to ill-defined parameters associated with pairing the correct ratio or combination of NPs with the drug of interest. Currently, research efforts are focused on trying to understand how to release diagnostic molecules and drugs from liposomes with heat, and microbubbles using ultrasound. Last but not the least, it will be important to understand how nano medicines behave when encountering different physiological characteristics of patients and their disease states.

Further Reading

Gupta PK. Fundamentals of nanotoxicology. 1st ed. New York: Elsevier; 2022.

Hirst KK. Lustreware – medieval Islamic pottery. ThoughtCo, August 27, 2020. https://www.thoughtco.com/what-is-lustreware-171559

Jaison J, Ahmed B, Chan YS, Alain D, Michael KD. Review on nanoparticles and nanostructured materials: history, sources, toxicity and regulations. Beilstein J Nanotechnol. 2018;9:1050–74. https://doi.org/10.3762/bjnano.9.98. www.ncbi.nlm.nih.gov/pmc/articles/PMC5905289/

Nikalje AP. Nanotechnology and its applications in medicine. Med Chem. 2015;5(2):081–9. https://doi.org/10.4172/2161-0444.1000247.

NNI. Nanotechnology timeline. National Nanotechnology Initiative; 2014. Official website of the United States. https://www.nano.gov/timeline#

Chapter 2
Sources, Classification, Synthesis, and Biomedical Applications

Abstract As per the National Academies, a distinction is made between three forms of nanoscale particles [usually shortened in the literature as "NPs" (nanoparticles) or "NMs"]—engineered, incidental, and natural. There is a need in particular for an overview of sources, classification, Synthesis, and biomedical applications and characterization of NMs. Two principal factors cause the properties of NMs to differ significantly from other materials: increased relative surface area, and quantum effects. These factors can change or enhance properties such as reactivity, strength, and electrical characteristics. NMs/NPs can be classified into different types according to the origin, size, morphology, dimension, physical and chemical properties, matrix involving geometry (particles, 1D forms, 2D forms), and chemistry (metals, semiconductors, ceramics, carbons, polymers). The interest in synthesis of NMs has grown because of their distinct optical, magnetic, electronic, mechanical, and chemical properties compared with those of the bulk materials. The fabrication and process are the key issues in nanoscience and nanotechnology to explore the novel properties and phenomena of NMs. Their potential applications in biomedicine and pharmaceuticals are very broad. Some important areas where nanotechnology has played a great role include: drug delivery, diagnostic techniques, antibacterial treatments, wound treatment, cell repair, tissue engineering, biosensors, and immunotherapy. In addition, synthesis of NPs and nanostructured materials (NSMs) represents an active area of research and a techno-economic sector with full expansion in many application domains.

Keywords Nanomaterials · Nanotechnology · Nanotoxicology · Sources · Properties · Classification · Synthesis · Biomedical applications · Natural NPs · Synthetic NPs · Nanosize · Engineered NMs · Incidental NMs · Bulk materials

© The Author(s), under exclusive license to Springer Nature Switzerland AG 2023   23
P K Gupta, *Nanotoxicology in Nanobiomedicine*,
https://doi.org/10.1007/978-3-031-24287-8_2

2.1   Introduction

Nanoparticles (NPs) and nanostructured materials (NSMs) are available from different sources having different properties. Biological systems often feature natural, functional nanomaterials (NMs). The structure of foraminifera (mainly chalk) and viruses (protein, capsid), the wax crystals covering a leaf, spider and spider-mite silk, the blue hue of tarantulas, the "spatula" on the bottom of gecko feet, some butterfly wing scales, natural colloids, horny materials and even our own bone matrix are all natural organic NMs. In addition, the synthesis of NPs and NSMs represents an active area of research and a techno-economic sector with full expansion in many application domains. This chapter briefly covers sources, classification, synthesis, characterization, properties, and biomedical application of NPs and NSMs.

2.2   Bulk Materials Versus NPs/NMs

Marked differences in properties and behaviors of bulk materials or larger particles vs. NPs/NMs classes are as follows:

Two principal factors cause the properties of NMs to differ significantly from other materials: increased relative surface area and quantum effects. These factors can change or enhance properties such as reactivity, strength, and electrical characteristics. As a particle decreases in size, a greater proportion of atoms are found at the surface compared to those inside. For example, a particle of size 30 nm has 5% of its atoms on its surface, at 10 nm 20% of its atoms, and at 3 nm 50% of its atoms. Thus NPs have a much greater surface area per unit mass compared with larger particles. As growth and catalytic chemical reactions occur at surfaces, this means that a given mass of material in the nanoparticulate form will be much more reactive than the same mass of material made up of larger particles.

2.3   Types

The main types of nanomaterial (NM) used in medical applications are semiconductor nanomaterials, magnetic nanomaterials, metal nanoparticles, carbon nanomaterials, hydrogel nanocomposites, liposomes, dendrimers, polymer nanocomposites, and biodegradable polymers. Nanoparticles (NPs) and nanostructured materials (NSMs) used in the medical field can be organized into four material-based categories (Table 2.1).

Nanoparticles can be synthesized from various organic or inorganic materials such as lipids, proteins, synthetic/natural polymers, and metals. Nanoparticles can be classified into several groups such as:

 (i) Polymeric nanoparticles, liposomes or lipid nanoparticles, dendrimers, and hydrogel

Table 2.1 Types of NPs and NSMs commonly used in the medical field

Carbon-based NMs	Generally, these NMs contain carbon and are found in morphologies such as hollow tubes, ellipsoids, or spheres. Fullerenes (C60), carbon nanotubes (CNTs), carbon nanofibers, carbon black, graphene (Gr), and carbon onions are included under the carbon-based NMs category. Laser ablation, arc discharge, and chemical vapor deposition (CVD) are the important production methods for these carbon-based materials' fabrication (except carbon black)
Inorganic-based NMs	These NMs include metal and metal oxide NPs and NSMs. These NMs can be synthesized into metals such as Au or Ag NPs, metal oxides such as TiO_2 and ZnO NPs, and semiconductors such as silicon and ceramics
Organic-based NMs	These include NMs made mostly from organic matter, excluding carbon-based or inorganic-based NMs. The utilization of noncovalent (weak) interactions for the self-assembly and design of molecules helps to transform the organic NMs into desired structures such as dendrimers, micelles, liposomes, and polymer NPs
Composite-based NMs	Composite NMs are multiphase NPs and NSMs with one phase on the nanoscale dimension that can either combine NPs with other NPs or NPs combined with larger or with bulk-type materials (e.g., hybrid nanofibers) or more complicated structures, such as a metal-organic framework. The composites may be any combination of carbon-based, metal-based, or organic-based NMs with any form of metal, ceramic, or polymer bulk materials

(ii) Inorganic nanoparticles, based on the components used for synthesis or the structural aspects of the NP

The fabrication methods and the properties of nanoparticles would also determine its application and utility. However, the type of nanoparticle used in the targeted delivery of therapeutics has its own positive and negative influences.

Various types of nanomaterials and their morphological features are summarized in Fig. 2.1.

2.4 Classification

Nanoparticles can be classified into different types according to their origin; size; morphology; dimension; physical and chemical properties; matrix involving geometry (particles, 1Dforms, 2D forms); and chemistry (metals, semiconductors, ceramics, carbons, polymers. Some of them are carbon-based nanoparticles, ceramic nanoparticles, metal nanoparticles, semiconductor nanoparticles, polymeric nanoparticles, and lipid-based nanoparticles). Some classifications distinguish between organic and inorganic nanoparticles; the first group includes dendrimers, liposomes, and polymeric nanoparticles, while the latter includes fullerenes, quantum dots, and gold nanoparticles.

Based on dimensions, NMs are classified as zero dimensional (0D), one dimensional (1D, Graphene, thin film), two dimensional (2D, carbon nanotubes), and three dimensional (3D, quantum dots or nanoparticles and fullerene) nanostructures. Examples are summarized in Table 2.2.

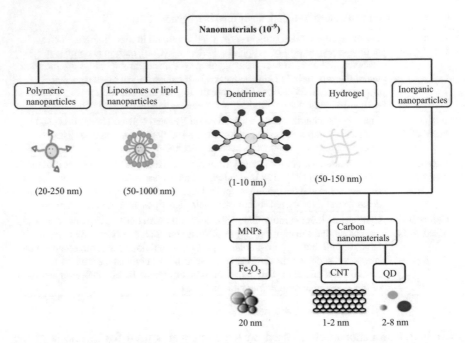

Fig. 2.1 Various types of NMs and their morphological features
https://media.springernature.com/full/springer-static/image/art%3A10.1186%2Fs40824-019-0166-x/MediaObjects/40824_2019_166_Fig 1_HTML.png?as=webp

Table 2.2 Different dimensions of NMs with examples

Elementary units	Examples
0D units (three dimensions in the monometric range)	Molecules, clusters, fullerenes, rings, metal carbides, powders, and grains
1D units (two dimensions in the monometric range)	Nanotubes, fibers, filaments, spirals, belts, whiskers, springs, columns, needles, etc.
2D units (one dimension in the monometric range)	Layers and asbestos

Hybrid Nanomaterials

Hybrid nanomaterials are defined as unique chemical conjugates of organic and/or inorganic materials that could be synthesized either by chemical or physical methods. That is, these are mixtures of two or more inorganic components, two or more organic components, or at least one of both types of components. The resulting material is not a simple mixture of its components but a synergistic material with properties and performance to develop applications with unique properties, which are determined by the interface of the components at the molecular or supramolecular level. Its functionality is associated with the improvement of physicochemical properties. Electrochemical or biochemical properties are mainly magnetic, electronic, optical, and thermal properties or a combination of them. The unique versatility of these materials allows designing materials with tunable properties, with improved performance and properties to their long-established counterparts in the market. The diversity of organic and inorganic components that can be incorporated into these materials is from sizes of a few angstroms to thousands of angstroms, so these can be categorized among molecular species, in nano- and/or supramolecular sizes, or with extended structure, for example, Donor-acceptor perovskites, intergrowth organic–inorganic perovskites, sol–gel silica modified with organic molecules, active organic molecules doped into conductive polymers, organically grafted inorganic phases, etc.

These materials can be categorized as under:

(a) First class of hybrid materials (those based on the synergy of the phases through weak chemical interactions based on Coulomb forces, London dispersion forces, hydrogen bonds, and dipole–dipole forces). Hybrid materials based on the first approach can be subdivided into two types, namely, (1) inorganic materials modified by organic moieties and (2) colloidal polymers stabilized by organic moieties.
(b) Second class of hybrid materials (those based on the synergy of the phases through strong chemical bonds such as Lewis acid–base, covalent, or ionic-covalent bonds). Hybrid materials based on the second approach can be subdivided into two types, namely, organic materials modified by inorganic constituents and biological organic materials modified by inorganic constituents.

A more recent classification of hybrid materials is based on their functionality. Three different types of materials can be identified:

(a) Structurally hybridized materials
(b) Functionally hybridized materials
(c) Hybridized materials in their chemical bond

2.5 Sources

Every person has been exposed to nanometer-sized foreign particles; we inhale them with every breath and consume them with every drink. In truth, every organism on Earth continuously encounters nanometer-sized entities. Sources of such nanomaterials can be classified into three main categories based on their origin:

 (i) Naturally produced nanomaterials, which can be found in the bodies of organisms, insects, plants, animals, and human bodies. However, the distinctions between naturally occurring, incidental, and manufactured NPs are often blurred. In some cases, for example, incidental NMs can be considered as a subcategory of natural NMs.
(ii) Incidental nanomaterials, which are produced incidentally as a byproduct of industrial processes such as nanoparticles produced from vehicle engine

exhaust, welding fumes, combustion processes, and even some natural processes such as forest fires.

(iii) Engineered nanomaterials, which have been manufactured by humans to have
certain required properties for desired applications.

Incidental and naturally occurring NMs are continuously being formed within
and distributed throughout ground and surface water, the oceans, continental soil,
and the atmosphere. One of the main differences between incidental and engineered
NMs is that the morphology of engineered NMs can usually be better controlled as
compared to incidental NMs; additionally, engineered NMs can be purposely
designed to exploit novel features that stem from their small size. It is known that
metal NPs may be spontaneously generated from synthetic objects, which implies
that humans have long been in direct contact with synthetic NMs and that mac-
roscale objects are also a potential source of incidental nanoparticles in the environment. Mostly, NPs used for commercial applications are engineered NPs that are
produced using physical, chemical, and biological methods.

2.6   Production Approaches

Fundamentally, there are two approaches to the production of NPs, the top-down
approach and the bottom-up approach. These approaches help to synthesize NMs of
desired size, shape, and orientation. The top-down approach is a process of minia-
turizing or breaking down bulk materials (macro-crystalline) structures while
retaining their original integrity. The bottom-up approach involves building nanomaterials from the atomic scale (assembling materials from atoms/molecules).
Synthesis of nanomaterials (NPs) via top-down and bottom-up approaches is shown
in Fig. 2.2.

(a) Top-Down Approach

The top-down approach involves the breaking down of the bulk material into
nanosized structures or particles that have been used for producing micron-sized
particles. This approach is inherently simpler and depends either on the removal or
division of bulk material or on the miniaturization of bulk fabrication processes to
produce the desired structure with appropriate properties. The biggest problem with
the top-down approach is the imperfection of surface structure. For example,
nanowires made by lithography are not smooth and may contain a lot of impurities
and structural defects on its surface.

(b) Bottom-Up Approach

The alternative approach, which has the potential of creating less waste and
hence the more economical, is the "bottom-up" This approach also refers to the
buildup of a material from the bottom: atom-by-atom, molecule-by-molecule, or
cluster-by-cluster. Many of these techniques are still under development or are just
beginning to be used for commercial production of nanopowders. Oraganometallic

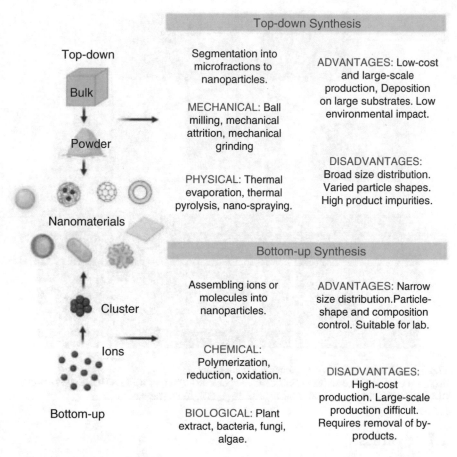

Fig. 2.2 Schematic representation of top-down and bottom-up techniques for nanomaterial fabrication
https://www.mdpi.com/journal/nanomaterials. Nanomaterials 2022, 12, 3226. https://doi.org/10.3390/nano12183226

Table 2.3 Differences in top-down and bottom-up implementation approaches

Bottom-up approach	Top-down approach
High deployment coverage in early phases	Tactical, limited coverage
Earlier return on investment	Delayed return on investment
High visibility of organizational changes	Lower impact on the overall organization
Higher impact on the organization	Higher deployment costs

chemical route, revere-micelle route, sol-gel synthesis, colloidal precipitation, hydrothermal synthesis, template-assisted sol-gel, electrodeposition, etc., are some of the well-known bottom-up techniques reported for the preparation of luminescent nanoparticles. Each approach has its advantages and disadvantages.

Differences in top-down and bottom-up implementation approaches have been summarized in Table 2.3.

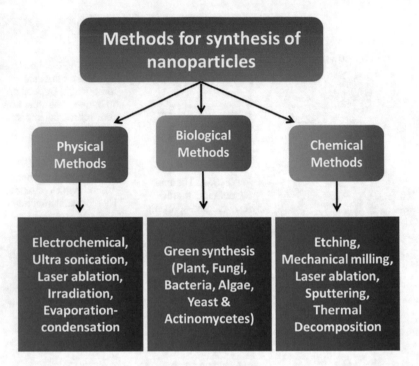

Fig. 2.3 Different methods for synthesis of nanoparticles
https://media.springernature.com/full/springer-static/image/art%3A10.1007%2Fs40097-018-0267-4/MediaObjects/40097_2018_267_Fig1_HTML.png?as=webp

2.7   Synthesis

The interest in the synthesis of NMs has grown because of their distinct optical, magnetic, electronic, mechanical, and chemical properties compared with those of the bulk materials. The fabrication and process are the key issues in nanoscience and nanotechnology to explore the novel properties and phenomena of NMs to realize their potential applications in science and technology. Many technological approaches/methods have been explored to fabricate NMs. NPs possess unique physicochemical, structural, and morphological characteristics, which are important in a wide variety of applications concomitant to electronic, optoelectronic, optical, electrochemical, environmental, and biomedical fields. Materials scientists and engineers have made significant developments in the improvement of methods of synthesis of NMs. In the past decade, there has been a marked increase in the field of fabrication of NPs with controlled morphologies and remarkable features, making it an extensive area of research. Other than their unique physical and chemical properties, NPs act as a bridge between bulk materials and atomic or molecular structures.

Physical, chemical, biological, and in some cases hybrid techniques are the main ways of NP production. The three main conventional methods of the synthesis of NPs are summarized in Fig. 2.3.

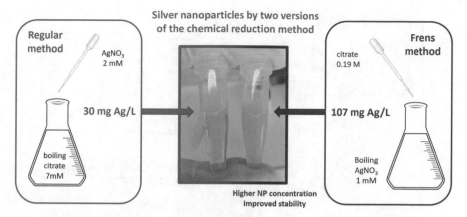

Fig. 2.4 Silver Nanoparticles by two versions of the chemical reduction method
https://www.mdpi.com/mps/mps-02-00003/article_deploy/html/images/mps-02-00003-ag.png

The physical method is a bottom-up approach to synthesize nanostructured materials, which involves two basic steps. The first step is the evaporation of the material, and the second step involves a rapidly controlled condensation to produce the required particle size. These physical methods of NP production include methods like laser ablation, high-energy irradiation, spray pyrolysis, and ion implantation. The physical NP production methods are mainly energy intensive and need special devices. For instance, the milling process is a way by which metallic microparticles are crushed using high-energy ball mills. The gas-phase process or aerosol process, which is divided into four main types (including flame reactor, plasma reactor, laser reactor and hot wall reactor, and chemical gas-phase deposition), is a particular way for the production of NPs like fullerenes and carbon nanotubes. All types of methods need special devices and are mainly high energy consuming.

The chemical methods include the reduction method (chemical reduction, indirect reduction) and the sedimentation method such as sol–gel process, co-precipitation, alkaline precipitation, and hydrothermal. Silver Nanoparticles by two versions of the chemical reduction method is shown in Fig. 2.4.

The chemical-mediated approach involves the use of various types of precursors, reducers, and stabilizers that has a unique feature because it allows an easy and independent control of the chemical compositions, structures, and morphologies of nanostructured materials and often allows the synthesis of NPs in large quantities. Moreover, the possibility of controlling particle size even at the nanometer scale is also possible during the chemical synthesis of NPs/NMs. NMs can be manufactured from metals, metal oxides, sulfides or selenides, carbon, polymers, or from biological molecules including lipids, carbohydrates, peptides, proteins, and nucleic acid oligomers.

The biological approach also called green NP biosynthesis involves the application of plant extracts; microorganisms (bacteria, fungi, algae, yeast, viruses, etc.); enzymes; and even some agricultural wastes for NP production. This approach for NP production is eco-friendly. The biologically produced NPs have special features

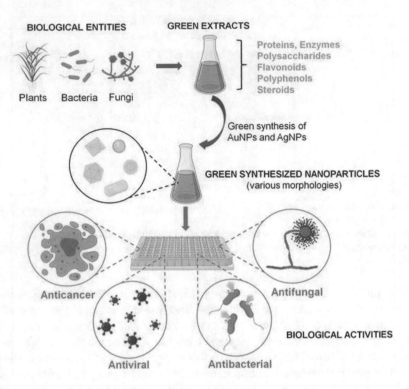

Fig. 2.5 Green, silver, and gold Nanoparticles: biological synthesis
https://www.mdpi.com/molecules/molecules-26-00844/article_deploy/html/images/molecules-26-00844-ag-550.jpg

including high catalytic activity, low toxicity contaminations, high stability, and plausible biocompatibility and biodegradability, making them distinctive from NPs produced from other methods. The microorganism's related NP productions may be intracellular synthesis or extracellular synthesis. (The intracellular method consists of transporting ions into the microbial cell to form NPs in the presence of enzymes. The extracellular synthesis of NPs involves trapping the metal ions on the surface of the cells and reducing ions in the presence of enzymes.) A mechanistic scheme with a graphical representation of the synthesis of metal NPs from microbes is presented in Fig. 2.5. The biological synthesis of NPs has a minimum toxic effect on the target body having a broad range of applications, such as targeted drug delivery, cancer treatment, gene therapy and DNA analysis, antibacterial agents, biosensors, enhanced reaction rates, separation science, and magnetic resonance imaging (MRI) over chemical synthesis. In the biological synthesis of metallic NPs, the use of hazardous chemicals, such as organic solvents and inorganic salts, is minimal which makes the process "green." Overall, the use of algae and waste materials for the green synthesis of NPs/NMs is an emerging and exciting area of nanotechnology and may have a significant impact on further advances in nanoscience.

2.8   Characterization

Nanoparticles are characterized for various purposes, including nanotoxicology studies and exposure assessment in workplaces to assess their health and safety hazards, as well as manufacturing process control. Characterization is sometimes incomplete without the use of reference material. This is because of the inherent difficulties of nanoscale materials to be properly analyzed, compared to bulk materials (e.g., too small size and low quantity in some cases following laboratory-scale production). In practice, and in agreement with the requirements mentioned above, characterization results should be obtained and used in their appropriate context scenarios. The information should be used for the description of intrinsic and extrinsic properties. The information on the properties provides a reliable basis as a starting point for the test and reference items used in such studies.

2.9   Properties

The main parameters of interest with respect to nanoparticle safety are the physical and chemical properties.

2.9.1   Physical Properties

Physical characteristics of nanoparticles and engineered nanomaterials include size, shape, specific surface area, aspect ratio, agglomeration/aggregation state, size distribution, surface morphology/topography, structure (including crystallinity and defect structure), and solubility.

2.9.2   Chemical Properties

Chemical properties include structural formula/molecular structure; composition of nanomaterial (including the degree of purity, known impurities or additives); phase identity; surface chemistry (composition, charge, tension, reactive sites, physical structure, photocatalytic properties, zeta potential); and hydrophilicity/lipophilicity.

When nanomaterials are used in test systems, one has to be aware that some of the properties which need to be determined are largely dependent on the surrounding media and the temporal evolution of the nanomaterials. Thus, a primary focus should be to assess the nanomaterials in exactly the form/composition they have as manufactured, and in the formulation delivered to the end-user or the environment if the formulation contains free nanoparticles.

Nanomaterials can exist as nanopowders: suspended in air (ultrafine particles, nanoparticles, aerosols), suspended in liquid (colloids), and incorporated in solids.

For biological safety evaluation, manufactured nanomaterials need to be dispersed in an appropriate media. The interaction between these media and the nanomaterials can have a profound influence on the behavior of the suspension.

The importance of the potential dissolution kinetics of newly manufactured nanomaterials needs to be emphasized. Since dissolution kinetics is frequently proportional to the surface area, nanomaterials are likely to dissolve much more rapidly than larger-sized materials. This applies, for example, to silver nanoparticles which are increasingly used for their release of silver ions as anti-bactericidal agents. Yet the dissolution kinetics is not properly studied. The example of silver nanoparticles highlights the complexity of risk estimates of nanomaterials since adverse interactions of the silver nanoparticles with biological systems need to be distinguished from those interactions of the ionic silver. However, not all properties can be determined in every situation, nor is it necessary to do so.

2.10 Applications

Nanomaterials have unusual mechanical, optical, electrical, and chemical behaviors, and they have been widely used in medicine and pharmaceuticals for the sensitive detection of key biological molecules, more precise and safer imaging of diseased tissues, and novel forms of therapeutics. The applications of nanomaterials in biomedicine and pharmaceuticals are very broad. Considering that this field is expanding rapidly, we cannot include all aspects of present nanomaterials in biomedicine and pharmaceuticals in detail. Some important areas where nanotechnology has played a great role include:

(a) Drug Delivery

Drug delivery involves employing nanoparticles to deliver drugs, heat, light, or other substances to specific types of cells (such as cancer cells). Particles are engineered so that they are attracted to diseased cells, which allows direct treatment of those cells. This technique reduces damage to healthy cells in the body and allows for earlier detection of disease.

(b) Diagnostic Techniques

Several diagnostic techniques in medicine where antibodies attached to carbon nanotubes in chips are used to detect cancer cells in the bloodstream. This method could be used in simple laboratory tests that could provide early detection of cancer cells in the bloodstream. Likewise, a test for early detection of kidney damage is being developed. The method uses gold nanorods functionalized to attach to the type of protein generated by damaged kidneys. When protein accumulates on the nanorod the color of the nanorod shifts. The test is designed to be done quickly and inexpensively for early detection of a problem.

(c) Antibacterial Treatments

Antibacterial treatments are being used for food preservation. The role of nanotechnology in various sectors of the food industry is summarized in Fig. 2.6. Other

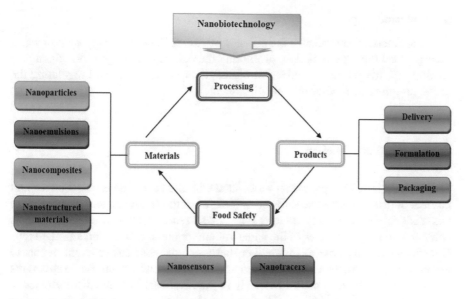

Fig. 2.6 Role of nanotechnology in various sectors of the food industry
https://www.mdpi.com/foods/foods-09-00148/article_deploy/html/images/foods-09-00148-g001.png

techniques are being developed to kill bacteria using gold nanoparticles and infrared light. This method may lead to improved cleaning of instruments in hospital settings. Another area is the use of quantum dots that can help to treat antibiotic-resistant infections.

(d) Wound Treatment

Skin wound healing is a major health care issue. On the bandage, an alternating discrete electric field is generated by a wearable nanogenerator by converting mechanical displacement from skin movements into electricity. Thus, these electrical pulses to a wound using electricity produced by nanogenerator can be used to treat the same.

(e) Cell Repair

Nanorobots could actually be programmed to repair specific diseased cells, functioning in a similar way to antibodies in our natural healing processes.

(f) Tissue Engineering

Nanoparticles have been used to serve various functions in tissue engineering, ranging from the enhancement of biological, electrical, and mechanical properties to gene delivery, DNA transfection, viral transduction, and patterning of cells, to facilitate the growth of various types of tissues to molecular detection and biosensing.

(g) Biosensors

Nanomaterials can be used for the immobilization of biomolecules; amplification of signals; and as mediators, electroactive species, and detection nanoprobes. Depending on the transduction mechanism, such as electrochemical, optical, and thermoelectric in biosensors, nanomaterials are being utilized.

(h) Immunotherapy

Nanomaterials can enhance the efficacy of cancer immunotherapy by protecting their payload during circulation, promoting the delivery of antigen-to-antigen presenting cell, triggering the activation of antigen-specific T cell, and regulating the immunosuppressive tumor microenvironment.

2.11   Current Status

The exploitation of the properties associated with the nanoscale is based on a small number of discrete differences between features of the nanoscale and those of more conventional sizes, namely the markedly increased surface area of nanoparticles compared to larger particles of the same volume or mass, and also quantum effects. Questions naturally arise as to whether these features pose any inherent threats to humans and the environment. There are several areas in which nanoscale structures are under active development or already in practical use. In biomedical science, a greater understanding of the functioning of molecules and the origin of diseases on the nanometer scale has led to improvements in drug design, drug delivery, cell and tissue repair, immunotherapy, and targeting. Nanomaterials are also being developed for analytical and instrumental applications, including tissue engineering and imaging. In medicine, a greater understanding of the origin of diseases on the nanometer scale is being derived, and drug delivery through functionalized nanostructures may result in improved pharmacokinetic and targeting properties.

Further Reading

Buzea C, Pacheco Blandino II, Robbie K. Nanomaterials and nanoparticles: sources and toxicity. Biointerphases. 2007;2(4):MR17–MR172.

Gupta PK. Fundamentals of nanotoxicology: concepts and applications. 1st ed. New York: Elsevier; 2022.

Harish V, Ansari MM, Tewari D, Gaur M, Yadav AB, García-Betancourt M-L, Abdel-Haleem FM, Bechelany M, Barhoum A. Nanoparticle and nanostructure synthesis and controlled growth methods. Nano. 2022;12:3226. https://doi.org/10.3390/nano12183226.

Jaison J, Ahmed B, Chan YS, Alain D, Michael KD. Review on nanoparticles and nanostructured materials: history, sources, toxicity and regulations. Beilstein J Nanotechnol. 2018;9:1050–74. www.ncbi.nlm.nih.gov/pmc/articles/PMC5905289/

Khanna P, Kaur A, Goyal D. Algae-based metallic nanoparticles: synthesis, characterization and applications. J Microbiol Methods. 2019;163(2019):105656. https://doi.org/10.1016/j.mimet.2019.105656.

Nikalje AP. Nanotechnology and its applications in medicine. Med Chem. 2015;5(2):081–9. https://doi.org/10.4172/2161-0444.1000247.

Shizhu C, Qun Z, Yingjian H, Jinchao Z, Xing-Jie L. Nanomaterials in medicine and pharmaceuticals: nanoscale materials developed with less toxicity and more efficacy. Eur J Nanomed. 2013;5(2):61–79. https://doi.org/10.1515/ejnm-2013-0003.

Warheit DB, Oberdörster G, Kane AB, Brown SC, Klaper RD, Hurt RH. Nanoparticle toxicology. In: Klaassen CD, editor. Casarett and Doull's toxicology. 9th ed. New York: McGraw-Hill Education. 2019; 1381–430.

Chapter 3
Mechanism of Nanotoxicity

Abstract The use of nanomaterials (NMs) has received much attention in the industrial and medical fields. Nanotoxicology is intended to address the toxicological activities of nanoparticles (NPs) and their products to determine whether and to what extent they may pose a threat to the environment and human health, and defined as the study of the nature and mechanism of toxic effects of nanoscale materials/particles on living organisms and other biological systems. This chapter consists of different sections with a brief introduction about the uptake, ADME (absorption, distribution, metabolism, and excretion) of NPs in vivo following different exposure routes. During the process of nanotoxicity, several major events such as cellular uptake pathways, binding to cell exterior, cell membrane interactions, intracellular trafficking, dissolution, involvement of reactive oxygen species (ROS), transcription of various pro-inflammatory genes, including tumor necrosis factor-α and IL (interleukins)-1, IL-6 and IL-8, by activating nuclear factor-kappa B (NF-κb) signaling, etc., are involved leading to the cascade of events of cellular injury and death.

Keywords NMs · Nanotoxicity · NPs · Toxicity · Oxidative stress · Omics · Mechanism of toxicity · Nanotechnology · Reactive oxygen species · Nano BioMedicine · Nano · Inflammation · Necrosis

3.1 Introduction

In the past decade, nano biomaterials have received considerable attention due to their wide variety of applications in medicine, which by themselves are not very harmful and could be toxic if they are ingested/inhaled or used in the form of NPs. Because of quantum size effects and large surface area to volume ratio, NMs have unique properties compared with their larger counterparts that affect their toxicity. Therefore, exposure to NMs may be dangerous, but the matter of their toxicity in

© The Author(s), under exclusive license to Springer Nature Switzerland AG 2023   37
P K Gupta, *Nanotoxicology in Nanobiomedicine*,
https://doi.org/10.1007/978-3-031-24287-8_3

humans is still unresolved. This chapter highlights the mechanism of nanomaterial (NM) toxicity including pathogenetic pathways activated by oxidative and non-oxidative stress leading to cell death. In addition, new insights on the complex molecular inter-relationships arising from "omics" in light of the information they can provide on specific intracellular events elicited by NMs have been discussed briefly.

3.2   Factors Affecting Nanotoxicity

The usage and development of NMs are increasing rapidly and are applied in various fields, leading to potential nanotoxicity in human beings and environmental risks. The toxicity of NMs is affected by their composition, much like the parent bulk materials. However, additional physicochemical properties play a crucial role in determining the toxicity of NMs, such as size, surface chemistry, shape, protein absorption gradient, and surface smoothness or roughness. Thus, the toxicity of chemically identical materials can be altered significantly by the manipulation of several physicochemical properties. The large number of variables influencing toxicity means that it is difficult to generalize about health risks associated with exposure to NMs—each new NM must be assessed individually and all material properties must be taken into account.

3.2.1   Physicochemical Factors

There has been a flurry of major advancement in the understanding of the interplay between particle size and shape for the development of a more efficacious NM-based targeted delivery system; nevertheless, this also reenforces that their untoward effects should also be examined. For example:

(a) Shape-dependent toxicity has been reported for myriads of NPs including carbon nanotubes, silica, allotropies, nickel, gold, and titanium. Likewise, wrapping processes in vivo during endocytosis or phagocytosis is dependent upon the shape of NPs. Rod-shaped SWCNT can block K+ ion channels two to three times more efficiently than spherical carbon fullerenes. In case of asbestos-induced toxicity, asbestos fibers longer than 10 μm caused lung carcinoma while fibers >5μm caused mesothelioma, and fibers >2μm caused asbestosis as longer fibers will not be effectively cleared from the respiratory tract due to the inability of macrophages to phagocytize them. NMs come in varied shapes, including fibers, rings, tubes, spheres, and planes, that can affect the toxicity (Fig. 3.1).

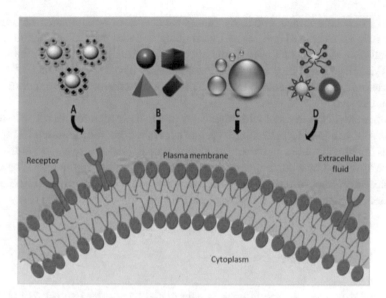

Fig. 3.1 Physicochemical factors that affect cellular uptake of NP
a = Surface charge, b = shape, c = size, and d = surface chemistry
https://media.springernature.com/full/springer-static/image/
art%3A10.1186%2Fs11671-018-2728-6/MediaObjects/11671_2018_2728_Fig2_HTML.
png?as=webp

(b) NPs show size-dependent toxicity, for example, Ag NPs with ≈10 nm diameter show a higher capacity to penetrate and disturb cellular systems of many organisms than Ag^+ ions and Ag NPs of larger diameters (20–100 nm).

(c) Reactivity and agglomeration of NPs are mostly dependent on their particle size. It is well known that the process of agglomeration will happen at slower rates in smaller particles than larger particles.

(d) Based on the crystal structure, NPs may exhibit different cellular uptake, oxidative mechanisms, and subcellular localization. For example, the two crystalline polymorphs of TiO_2 (rutile and anatase) show different toxicity. In the dark, rutile NPs (200 nm) lead to DNA damage via oxidation, while anatase NPs (200 nm) do not induce DNA damage in dark conditions.

(e) The surface properties of NPs have shown drastic effects relating to translocation and subsequent oxidation processes.

(f) Aggregation appears to be a ubiquitous phenomenon among all NPs and its influence in mediating cellular uptake and interactions is possible. Aggregation concentration and solubility of NPs at different concentrations can have an adverse effect on the toxicity. Increasing the NP concentration promotes aggregation so that a significant quantity of aggregated NPs may not penetrate cells thereby losing their toxicity. Weakly bound (agglomeration) and fused particles are significant risk criteria as they lead to poor corrosion resistance.

(g) NPs can be charged either by functionalization or spontaneous degradative reactions. Chemical species and their charge-related critical functional groups will be a significant factor in the specific functionality and bioavailability of NMs. However, there is no general consensus about the influence of charge (cationic, anionic, neutral) on the toxicity of engineered NMs on culture cells and in vivo.

(h) Inherently, NPs interact with impurities due to their high reactivity. Due to this reason, encapsulation becomes a prime necessity for solution-based NP synthesis (chemical route). In the encapsulation process, the reactive nano-entities are encapsulated by nonreactive species to provide stability to the NPs. For example, sulfur impurities may present in iron oxide NPs depending on the precursor used for their production ($FeCl_3$ or $Fe_2(SO_4)_3$).

3.2.2   Pre-exposure

The cellular phagocytic activity can be stimulated by shorter exposure time or the pre-exposure of lower NP concentrations. This pre-exposure results in the adaptability of the human body against NPs to some degree, and thereby may change the toxic potential of NP/NM.

3.2.3   Dose and Exposure Time

The number of NMs that penetrate the cells directly depends on the portal of entry and on the molar concentration of NPs in the adjacent medium multiplied by the exposure time.

3.2.4   Environmental Factors

Several environmental factors (such as temperature, pH, ionic strength, salinity, and organic matter) collectively influence NP behavior, fate and transport, and ultimately toxicity. The literature also seems insufficient to study the interactive behavior of factors affecting this phenomenon.

3.2.5   Other Factors

Other factors such as the chemical composition of NPs also affect the toxicity of nanoparticles.

3.3   Entry Sites and Uptake

NPs can enter the body directly through body openings, for example, inhaling, swallowing or through drug delivery, tissue implants (tissue engineering), or through any other mechanism. There is also an ongoing discussion about a potential indirect uptake through the skin pores. The human skin, the gastrointestinal tract, and the lungs are always in direct contact with the environment. Whereas the skin serves as a barrier, the gastrointestinal tract and lungs allow the (active or passive) transport of various substances such as water, nutrients, or oxygen. It seems likely that NPs can also enter the human body via these routes. Due to their small size, such particles can penetrate the cell membrane and show activity at the subcellular level.

The major route of exposure to NPs include (a) respiratory (inhalation exposure), (b) oral (digestive system), (c) dermal/skin, and (d) parental administration.

3.4   Fate in the Body

The extremely small size of NMs can much more readily enter the human body than larger-sized particles. NPs enter the body by crossing one of its outer layers, either the skin, the lining of the lungs, or the intestine. How well they transfer from outside to inside will depend on the particular physical and chemical properties of the particle. How these NPs behave inside the body is still a major question that needs to be resolved. The behavior of NPs is a function of their size, shape, and surface reactivity with the surrounding tissue. In principle, a large number of particles could overload the body's phagocytes, cells that ingest and destroy foreign matter, thereby triggering stress reactions that lead to inflammation and weaken the body's defense against other pathogens. In addition to questions about what happens if non-degradable or slowly degradable NPs accumulate in bodily organs, another concern is their potential interaction or interference with biological processes inside the body. Because of their large surface area, NPs will, on exposure to tissue and fluids, immediately adsorb onto their surface some of the macromolecules they encounter.

NMs cross the biological membrane either through simple diffusion or passive transport.

NPs and NMs enter the vascular system and are distributed to the organs and peripheral tissues of the body. Within the vascular compartment NPs encounter blood cells, platelets, coagulation factors, and plasma proteins and depending on the size and charge, NPs and NMs may undergo adsorption or opsonization by serum proteins. The vascular endothelial cell monolayer acts as a dynamic, semiselective barrier that regulates the transport of fluid and macromolecules between the vascular compartment and the extravascular space. Although the structure of the endothelial layer varies throughout the body, the effective pore size in normal intact endothelium is about 5 nm. Nano-sized molecules with a hydrodynamic diameter (HD) less than 5 nm achieve rapid equilibrium with the extravascular extracellular space (EES), whereas larger particles experience prolonged circulatory times due to slow transport across the endothelium.

Then, once in the bloodstream, NMs can be transported around the body and be taken up by organs and tissues, including the brain, heart, liver, kidneys, spleen, bone marrow, and nervous system. NMs can be toxic to human tissue and cell cultures (resulting in increased oxidative stress, inflammatory cytokine production, and cell death) depending on their composition and concentration.

The excretion of NPs and NMs from the body is mainly through feces and urine. Urinary excretion is restricted to nanostructures <5.5 nm in size for metal-based NP, limited by the hydrodynamic diameter of the particle that may change in circulation due to protein adsorption. This general principle appears to be different for circulating fibrous structures of NM. The fecal excretory clearance pathways consist of several inputs: One is mucociliary clearance of deposited particles from the airways into the GI tract; another is via hepatobiliary clearance of blood-borne NM via the liver and bile into the small intestine. This elimination pathway is also a well-known excretory path for heavy metals in the blood.

With regard to the CNS, no data on NM elimination are available yet. It is conceivable that the CSF via its connections to the nasal lymphatic system and to blood circulation could be an excretory pathway for the brain, which needs to be investigated.

3.5 Cellular Uptake Pathways

Nanoscale materials are increasingly found in consumer goods, electronics, and pharmaceuticals. While these particles interact with the body in myriad ways, their beneficial and/or deleterious effects ultimately arise from interactions at the cellular and subcellular levels.

NPs can modulate (a) cell fate, (b) induce or prevent mutations, (c) initiate cell-cell communication, and (d) modulate cell structure in a manner dictated largely by phenomena at the nano-bio interface. The cell membrane protects intracellular components from the surrounding environment. More specifically, the cell membrane maintains (i) cell homeostasis, (ii) provides structural support, (iii) maintains ion concentration gradients, and (iv) controls the entry and exit of charged small molecules and nutrients. Almost all natural membranes, regardless of function, share a common general structure: a bilayer of amphiphilic lipids with hydrophilic heads and hydrophobic tails. The amphiphilic properties of phospholipids make their bilayer assembly an efficient selective barrier, as "balanced" hydrophobicity/hydrophilicity is needed to permit a wide range of small biomolecules to enter the cell by passive diffusion. However, entry is regulated in some cases through other mechanisms (e.g., channel, receptor, or transporter).

NPs acquire different physicochemical properties in biological fluids such as blood and cell-culture media, therefore, the surface of NPs is dramatically modified by the adsorption of biomolecules including proteins, the so-called protein corona. Thus, the response of NP varies depending upon various factors such as:

(a) NP-related factors including the collective physicochemical properties of NPs such as shape and size, polydispersity, charge, surface chemistry, and surface hydrophobicity/hydrophilicity (Fig. 3.1).

Before the NPs reach the exterior membranes of target cells, they must interact with the microenvironment around the target cells. Furthermore, that microenvironment, including fibrosis, extracellular matrix, various microenvironmental factors, pH and so on, can also change the properties of NPs and affect their interactions with the cell membrane and finally their intracellular fate.

However, NPs designed for targeting tumor cells behave differently because the tumor microenvironment could have a great influence on their cellular fate. Therefore, NPs used for different applications may behave differently in the microenvironment of the target cell because it greatly influences the performance of NPs, determining where they go and what kind of cells they interact with.

3.6 Cell Membrane Interactions

When NPs reach the exterior membrane of a cell, they can interact with components of the plasma membrane or extracellular matrix and enter the cell, mainly through endocytosis or through other mechanisms.

3.6.1 Endocytosis

Endocytosis is a form of active transport. This process leads to the engulfment of NPs in membrane invaginations, followed by their budding and pinching off to form endocytic vesicles, which are then transported to specialized intracellular sorting/trafficking compartments. Depending on the cell type, as well as the proteins, lipids, and other molecules involved in the process, endocytosis can be classified into several types. The different types of endocytosis and exocytosis patterns of NPs are shown in Fig. 3.2. The main mechanisms of endocytosis include:

(a) Phagocytosis (clathrin-mediated endocytosis, caveolin-mediated endocytosis, and clathrin/caveolae-independent endocytosis are subtypes of the broadly defined process of phagocytosis)
(b) Micropinocytosis

Compared to phagocytosis, which takes place mainly in professional phagocytes, pinocytotic mechanisms are more common and occur in many cell types.

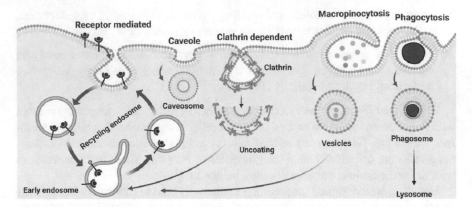

Fig. 3.2 Different pathways of endocytosis
https://pub.mdpi-res.com/ijms/ijms-24-02971/article_deploy/html/images/ijms-24-02971-g001.
png?1675407450

3.6.2 Other Entry Mechanisms

Other entry mechanisms include passive diffusion, hole formation, direct microinjection, and electroporation. Several nano-sized materials can cross cellular membranes by passive pathways such as diffusion and adhesive interactions, where thermal capillary waves and line tension play a significant role in controlling the entry of NPs into cells. A range of cationic NPs made of both organic (peptides and polycationic polymers) and inorganic (Au-NH2, SiO2-NH2) materials can penetrate membranes by disruption of lipid bilayers and nanoscale hole formation.

3.7 Intracellular Trafficking

Following uptake, the next crucial matter is the intracellular trafficking of NPs which determines its final destination within cellular compartments, its cytotoxicity, and its therapeutic efficacy. After NPs are internalized by the cells, they will first encounter membrane-bound intracellular vesicles called early endosomes. Endosomes formed at the plasma membrane are categorized into three types: early endosomes, late endosomes, and recycling endosomes (Fig. 3.2). NPs in the cytoplasm or trapped in vesicles can enter the nucleus, mitochondria, endoplasmic reticulum (ER), and Golgi apparatus via unknown mechanisms. In fact, vesicles containing NPs can fuse with ER, Golgi, and other organelles. NPs that enter the ER or Golgi may leave the cell via vesicles related to the conventional secretion system. NPs that are localized in the cytoplasm can leave the cells by reentering the vesicular system or directly via unspecific mechanisms. During the intracellular fate of NPs, there may be a process of autophagy (autophagy is an intracellular degradation pathway that, distinct from the endocytic pathway, delivers certain cytoplasmic constituents to lysosomal degradation).

3.8   Cellular Exocytosis

During cellular exocytosis materials from within a cell move to the exterior of the cell. Exocytosis of NPs (Fig. 3.2) involves vital biological processes such as organizing membrane proteins (e.g., transporters, ion channels, and receptors) and lipids; excretion of essential molecules; and repairing the cell membrane. After cellular exocytosis, NPs have been shown to acquire the ability to cross critical in vivo barriers, such as the blood-brain barrier, and cause unexpected cytotoxicity. Exosomes carrying inhaled NPs out of alveolar cells and disseminating them into the systemic circulation have also been shown to induce systemic immune responses and subsequent inflammation; this further emphasizes the role of NPs' exocytosis in their potential toxicity in the human body.

3.9   Processes of Nanotoxicity

The main processes of nanotoxicity include (i) binding to the cell exterior, (ii) dissolution, and (iii) ROS. Depending on the type of cell, NPs can also enter the cell after binding and potentially cause toxicity.

3.9.1   Binding to Cell Exterior

All cells generate an electrical potential across their plasma membrane driven by a concentration gradient of charged ions. A typical resting membrane potential ranges from -40 to -70 mV, with a net negative charge on the cytosolic side of the membrane. One of the biocidal mechanisms induced by these NPs is direct NPs sticking to the surface of a cell can disrupt the cell's membrane, which can kill the cell. A schematic representation of the antimicrobial mechanisms of NPs is shown in Fig. 3.3. This mechanism is a creative and efficient way to prevent microbial growth in foods, coupled with the preservation of their quality, freshness, and safety. Packaging with the aim of preserving the quality of food systems is considered a solution to overcome this challenge. NPs having large surface-to-volume ratio provide more direct interaction with bacterial surfaces, and these NMs show excellent antibacterial properties. Particularly, cationic NPs are firmly attached to the membrane of bacteria with negatively charged outer layers by electrostatic interactions. Disruption of the cell integrity results in the leakage of cell contents. NPs have intrinsic antibacterial activities to refuse the microbes by mimicking the natural course of killing by phagocytic cells, that is, by producing a large quantity of

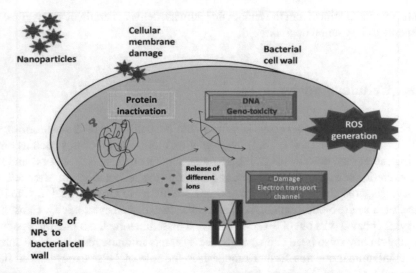

Fig. 3.3 Schematic representation of antimicrobial mechanisms of NPs
https://media.springernature.com/lw685/springer-static/image/art%3A10.1007%2Fs10904-020-01674-8/MediaObjects/10904_2020_1674_Fig1_HTML.png?as=webp

reactive nitrogen species (RNS) and ROS. Besides, NMs could also prevent or overcome biofilm formation. NPs especially metallic NPs exert toxic effects by enhancing natural immunity or mimicking natural immune responses by generating a large quantity of RNS or ROS. Others possibly exert direct killing effects may be by directly targeting cellular proteins, DNA, or lipids. Thus, the mechanism underlying the direct killing effects may be attributed to:

1. Electrostatic interactions between cell walls and NPs to destroy bacterial cell integrity.
2. Liberation of antimicrobial ion accumulation of NPs into bacteria cells.
3. ROS formation and possible change in bacterial surface properties, such as distortion caused by the leakage of intracellular contents after NPs treatment.

Accordingly, antimicrobial packaging in the food sector is a robust nanoscale technology to fulfill the aforementioned requirements.

Another important aspect of this type of mechanism of nanotoxicity helps in several options for redesigning a NP to reduce its ability to bind the cell wall. One involves changing the NP's surface charge. Most cells have a negative charge at their cell surface, and it is well known that like-repels-like (just like with magnets). So, if NPs are designed to have negative surface charges, they will be less capable of binding to a cell surface and therefore cannot damage cells due to binding.

3.9.2 Dissolution

The second way NPs can be toxic to organisms happens when they dissolve in liquid, which releases ions (charged molecules) into the environment. These ions can interact with and be taken up by cells, sometimes with harmful effects. Another way to redesign NPs to reduce the release of toxic ions is to replace the elements that are toxic in their ionic form with something less toxic.

3.9.3 Reactive Oxygen Species

An important mechanism of nanotoxicity is the generation of ROS. Overproduction of ROS can induce oxidative stress, resulting in cells failing to maintain normal physiological redox-regulated functions. Both in vivo and in vitro studies have shown that NPs are closely associated with toxicity by increasing intracellular ROS levels and/or the levels of pro-inflammatory mediators. The homeostatic redox state of the host becomes disrupted upon ROS induction by NPs. In excess it has been found to cause severe damage to cellular macromolecules such as proteins, lipids, and DNA, resulting in detrimental effects on cells. Several studies with differently sized cerium oxide, silver, gold, and silica multi-walled carbon nanotubes NPs indicated that they exert their toxicity through oxidative stress. ROS generation decreased mitochondrial membrane potential, increased levels of lipid peroxide, and decreased enzymatic activities of antioxidants (Fig. 3.4). If the extent of DNA damage goes beyond the scope of repair by the DNA repair mechanisms of the body, the cells initiate programmed cell death.

In order to reduce ROS generation, NPs can be coated with a shell made of a material that is less prone to ROS production. For example, adding a zinc sulfide shell to cadmium selenide quantum dots reduce their production of ROS without changing the functionality of the quantum dots. This demonstrates that coating can be an effective method of reducing impacts on organisms due to ROS generation (Fig. 3.4).

3.9.4 Inflammation-Mediated Nanotoxicity

Inflammation is a defense mechanism of the body that involves several immune regulatory molecules, following the infiltration of phagocytic cells. NPs such as single and multi-walled carbon nanotubes (MWCNT) and fullerene derivatives are also known to upregulate the transcription of various pro-inflammatory genes, including tumor necrosis factor-α and IL (interleukins)-1, IL-6 and IL-8, by activating nuclear factor-kappa B (NF-κB) signaling (Fig. 3.4). The computational model suggests that the carbon nanotubes (CNT) and C60 fullerenes may be recognized as pathogens by the Toll-like receptors, triggering innate immune responses of the

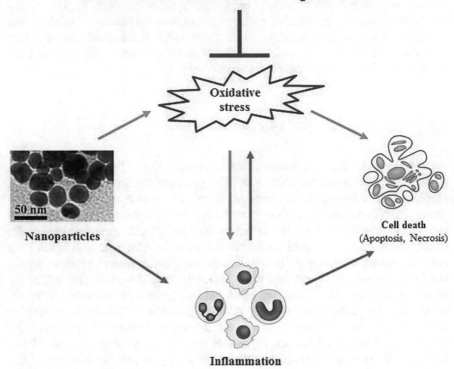

Fig. 3.4 Overview-of-the-signaling-cascades-mediating-nanotoxicity-inflammation and cell death
https://pub.mdpi-res.com/nanomaterials/nanomaterials-05-01163/article_deploy/html/images/
nanomaterials-05-01163-ag-550.jpg?1581016419

body and secretion of inflammatory protein mediators such as interleukins and chemokines. Furthermore, activation of the complement cascade on exposure to liposomes and other lipid-based NPs leads to hypersensitivity reactions and anaphylaxis.
The exact mechanism of how these complement proteins mediate nanotoxicity has
not yet been elucidated. However, these sequential molecular and cellular events are
known to cause oxidative stress, followed by severe cellular genotoxicity and then
programmed cell death.

3.10 Pathways of Cellular Dysfunction

It is now well known that NPs can cause adverse effects on organelles, which have
implications on mitochondrial dysfunction, endoplasmic reticulum stress, and lysosomal rupture. There is initially a sublethal phase in the pathogenesis of cell injury

from which a cell can recover or, alternatively, cells may pass a "point of no return" (permeabilization of the mitochondrial membranes). The point of no return is associated with a crucial separation between the outer and inner mitochondrial membranes followed by the irreversible loss of oxidative phosphorylation capacity. Finally, in the late stages of necrosis, the cytoplasm loses contents and takes on a homogeneous eosinophilic appearance (as with ground glass), irregularities in the membrane of cytoplasmic organelles, mitochondrial swelling, increased matrix density, the formation of vacuoles, and the deposit of calcium phosphates. At the nuclear level, chromatin patterns are seen with pyknosis (chromatic condensation), karyorrhexis (nuclear fragmentation) and karyolysis (complete chromatin disruption), the disturbance of mitochondrial biogenesis and mitochondrial dynamic fusion-fission, mitophagy, and cytochrome c-dependent apoptosis are involved. In addition, prolonged endoplasmic reticulum stress will result in apoptosis. NPs do not share a common mode of action, and both the type of cell damage (e.g., necrosis, apoptosis, inhibition of proliferation) and the underlying mechanism for the same type of cell damage show particle-related differences (e.g., induction of apoptosis by destabilization of mitochondria or by disruption of lysosomes). A schematic representation of signaling cascade during nanotoxicity that can lead to inflammation, cellular dysfunction, apoptosis, necrosis, or cell death is shown in Fig. 3.4.

The most common cell dysfunction includes:

- Damage to the cell wall and plasma membrane
- Interference with electron transport and aerobic respiration
- Induction of oxidant stress
- Activation of cell signaling pathways
- Perturbed ion homeostasis
- Release of toxic metal ions from internalized NPs
- Disruption of lysosomal membrane integrity
- Incomplete uptake or frustrated phagocytosis
- Interference with cytoskeletal function
- DNA and chromosomal damage

Damage to the cell wall and plasma membrane is a highly complex and tightly regulated pathway involving several signaling molecules. As indicated earlier, the cell displays a series of changes in the early (reversible) stage that includes "change of hydropic," "degeneration of feathery," "cloudy swelling," or "vacuolar degeneration." The point of no return is associated with a crucial separation between the outer and inner mitochondrial membranes followed by the irreversible loss of oxidative phosphorylation capacity. Finally, in the late stages of necrosis, the cytoplasm loses contents and takes on a homogeneous eosinophilic appearance (as with ground glass), irregularities in the membrane of cytoplasmic organelles, mitochondrial swelling, increased matrix density, the formation of vacuoles, and the deposit of calcium phosphates. At the nuclear level, chromatin patterns are seen with pyknosis (chromatic condensation), karyorrhexis (nuclear fragmentation), and karyolysis (complete chromatin disruption).

3.10.1   Apoptosis Versus Necrosis

Apoptosis is a form of cellular suicide that can be classified into extrinsic and intrinsic apoptosis. NMs are described as triggers of extrinsic and intrinsic apoptotic pathways. Extrinsic apoptosis indicates cell death, caspase-dependent, stimulated by extracellular stress signals that are sensed and propagated by specific transmembrane receptors. Intrinsic apoptosis can be triggered by a plethora of intracellular stress conditions, such as DNA damage, oxidative stress, and many others. It results from a bioenergetic and metabolic catastrophe coupled with multiple active executioner mechanisms. Although the mode of cell death depends on the severity of the cellular insult (which may, in turn, be linked to mitochondrial function and intracellular energy), it has been difficult to set up a comprehensive mechanism of NMs cell death based on conflicting observations present in the literature. Metal oxide NPs including TiO_2, ZnO, Fe_3O_4, Al_2O_3, and CrO_3 of particle sizes ranging from 30 to 45 nm are found to induce apoptosis. Despite the extensive use of NPs today, there is still a limited understanding of NP-mediated toxicity. In the literature, there are confused and inconsistent examples of necrosis induced by NMs, because on one hand only the loss of cell viability is often evaluated without focalizing on the cell death modalities, and on the other hand, there are no single discriminative biochemical markers available yet. However, a recent study demonstrated that water-soluble germanium NPs with allylamine-conjugated surfaces (4 nm) induce necrotic cell death that is not inhibited by necrostatin-1 in Chinese hamster ovary cells. Different steps involved during the process of necrosis and apoptosis are summarized in Fig. 3.5.

More and accurate results are needed for apoptosis, necrosis, and autophagy or "autophagic cell death induction" by NMs; further studies are necessary to test if the novel strategic targets identified could be affected either directly or indirectly by NMs. However, the major differences between necrosis and apoptosis have been summarized in Table 3.1.

3.10.2   Autophagy or "Autophagic Cell Death"

Autophagic cell death (ACD) is morphologically defined (especially by transmission electron microscopy) as a type of cell death that occurs in the absence of chromatin condensation but is accompanied by large-scale autophagic vacuolization of the cytoplasm. Several classes of NMs such as alumina, europium oxide, gadolinium oxide, gold, iron oxide, manganese, neodymium oxide, palladium, samarium oxide, silica, terbium oxide, titanium dioxide, ytterbium oxide, and yttrium oxide NPs; nanoscale carbon black; fullerene and fullerene derivate; and protein-coated quantum dots induce elevated levels of autophagic vacuoles in different animals and human cell culture as well as in vivo models. NMs may induce autophagy via an oxidative stress mechanism, such as the accumulation of damaged proteins and

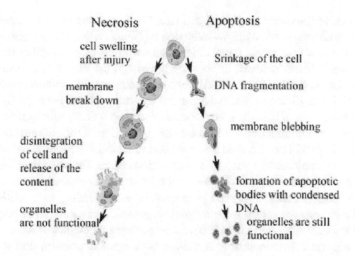

Fig. 3.5 Different steps involved during the process of necrosis and apoptosis. Schematic representation of necrotic and apoptotic cells. In the figure are resumed the most evident modifications, induced by the two different processes, occurring in the cell

https://www.researchgate.net/profile/Adrianna-Ianora/publication/226221124/figure/fig1/
AS:669988553777164@1536749127534/Schematic-representation-of-necrotic-and-apoptotic-
cells-In-the-figure-are-resumed-the.png

Source: Buttino, J.-S. Hwang, C.-K. Sun et al., 2011. Apoptosis to predict copepod mortality: State of the art and future perspectives. Hydrobiologia 666(1):257–264 DOI: https://doi.org/10.1007/s10750-010-0536-9

Table 3.1 Differences between necrosis and apoptosis

Necrosis	Apoptosis
Cellular swelling	Cell shrinkage
Membranes are broken	Membranes remain intact
Cell lysis, eliciting an inflammatory reaction	Cell is phagocytosed, no tissue reaction
DNA fragmentation is random, pyknosia	DNA fragmentation into nucleosome size fragments
Mechanism—ATP depletion, membrane injury, free radical damage	Mechanism-caspase activation, endonuclease, and proteases
In vivo, whole areas of the tissue are affected	In vivo, individual cells appear affected

subsequent endoplasmic reticulum or mitochondrial stress and altering gene/protein expression and/or regulation, and interfering with the kinase-mediated regulatory cascades.

3.11 Conclusion

In spite of the recent advances in our understanding of cell death mechanisms and associated signaling networks, much work remains to be done before we can fully elucidate the toxicological behavior of the NMs as well as understand their participation in the determination of cell fate. More accurate results are needed for apoptosis, autophagy, and necrosis induction by NMs; further studies are necessary to test if the novel strategic targets identified could be affected either directly or indirectly by NMs. Moreover, no data are present in the literature concerning the NMs exposure and other forms of cell death, including anoikis, entosis, parthanatos, netosis, and cornification. Although numerous studies have been performed on keratinocytes, none of these has rated cornification, a cell death subroutine restricted to keratinocytes and functionally linked to the generation of the stratum corneum of the epidermis. It will be of considerable interest to establish various cell death delineating the mechanisms by which these interactions occur. Despite intensive investigations, the understanding of NMs-induced cellular damage remains to be clarified. However, in most cases, enhanced generation of ROS, leading to oxidative stress which in turn may trigger proinflammatory responses, is assumed to be responsible for NMs toxicity, although nonoxidative stress-related mechanisms have also been recently reported. In conclusion, a multilevel-integrated uniform and consistent approach should contemplate for NM toxicity characterization.

Further Reading

Ahmet A, Hande S, Mohammad C. Chapter 18: NPs toxicity and their routes of exposures. In: Recent advances in novel drug carrier systems. Intech; 2012. p. 483–500. https://doi.org/10.5772/51230.

Asha R, Hande, PV, Prakash M, Valiyaveettil, S. Anti-proliferative activity of silver NPs. BMC Cell Biol. 2009;10:65, pp. 1–14. https://doi.org/10.1186/1471-2121-10-65. http://www.biomedcentral.com/1471-2121/10/65

Buttino I, Hwang J-S, Sun C-K, et al. Apoptosis to predict copepod mortality: state of the art and future perspectives. Hydrobiologia. 2011;666(1):257–64. https://doi.org/10.1007/s10750-010-0536-9.

Eleonore F. Cellular targets and mechanisms in the cytotoxic action of non-biodegradable engineered NPs. Curr Drug Metab. 2013;14(9):976–88. www.ncbi.nlm.nih.gov/pmc/articles/PMC3822521/

Gupta PK. Chapter 15: Toxic effects of nanoparticles. In: Toxicology: resource for self study questions. 2nd ed. Seattle: Kinder Direct Publications; 2020a.

Gupta PK. Chapter 14: Toxicology of nanomaterial particles. In: Problem solving questions in toxicology – a study guide for the board and other examinations. 1st ed. Cham: Springer Nature; 2020b.

Gupta PK. Toxic effects of nanoparticles. In: Brain storming questions in toxicology. 1st ed. Boca Raton: Taylor & Francis Group, LLC/CRC Press; 2020c. p. 297–300.

Gupta PK. Fundamentals of nanotoxicology. 1st ed. New York: Elsevier Inc.; 2022.

Jaison J, Ahmed B, Chan YS, Alain D, et al. Review on NPs and nanostructured materials: history, sources, toxicity and regulations. Beilstein J Nanotechnol. 2018;9:1050–74. Published online 2018 Apr 3. https://doi.org/10.3762/bjnano.9.98.

Khanna P, Cynthia O, Boon HB, Gyeong HB. Nanotoxicity: an interplay of oxidative stress, inflammation and cell death. Nanomaterials. 2015;5(3):1163–80. https://doi.org/10.3390/nano5031163.

Mahmoud H, Seid MJ, Katouzian I. Inorganic and metal NPs and their antimicrobial activity in food packaging applications. Crit Rev Microbiol. 2018;44(2):161–81. https://doi.org/10.108 0/1040841X.2017.1332001.

McShan D, Ray PC, Yu H. Molecular toxicity mechanism of nanosilver. J Food Drug Anal. 2014;22(1):116–27.

Shahed B, Vahid SB, Tao W, Hamaly MA, Mahmoud Y, et al. Cellular uptake of NPs: journey inside the cell. Chem Soc Rev. 2017 July 17;46(14):4218–44. https://doi.org/10.1039/c6cs00636. https://www.intechopen.com/books/recent-advances-in-novel-drug-carrier-systems/ NPs-toxicity-and-their-routes-of-exposures

Walters C, Edmund P, Vernon S. Chapter 3: Nanotoxicology: a review. In: Toxicology – new aspects to this scientific conundrum. 2018. Conact.book@intechopen.com/, p. 45–63. https:// doi.org/10.5772/64754.

Warheit DB, Oberdörster G, Kane AB, et al. Nanoparticle toxicology. In: Klaassen CD, editor. Casarett and Doull's toxicology: the basic science of poisons. 9th ed. New York: McGraw-Hill Education; 2019. p. 1381–430.

Chapter 4
Organ and Non-organ-Directed Nanotoxicity

Abstract Humans are exposed to various nanoscale materials since childhood, and as such have become a threat to human life. Because of their small size, Nanoparticles (NPs) find their way easily to enter the human body and cross the various biological barriers and may reach the most sensitive organs. On the basis of available experimental models, it may be difficult to judge their toxicity. NPs may interfere with the normal physiological mechanisms of the embryos, growing animals, and adults, and it is indispensable to understand their potentially direct or indirect harmful effects on living organisms including human beings. The interaction between NPs and cell triggers a cascade of molecular events which could induce toxicity and cell death. Although, no mechanisms unique to NPs have yet been identified, however, some NPs are known to induce inflammatory changes, induction of reactive oxygen species, and the consequential oxidative stress in different cells of various tissues. This chapter briefly describes the toxicity of NPs vs. larger particles, influence of various factors, and organ and non-organ-directed toxicity in various tissues.

Keywords Exposure · Nanoparticles · Nanomaterials · Cytotoxicity · Metal nanoparticles · Toxicology · Organ nanotoxicity · Genotoxicity · Carcinogenicity · Inhalation toxicity · Airborne toxicity · Reactive oxygen species · Oxidative stress · Immune system

4.1 Introduction

Nanoparticles (NPs) fascinate medical scientists and engineers because they possess properties different from those seen in bulk samples of the same material. Humans are exposed to various nanoscale materials since childhood, and as such have become a threat to human life. In addition, NPs/NMs (nanomaterials) have emerged as important players in modern medicine, with clinical applications ranging from contrast agents in imaging to carriers for drug and gene delivery into tumors. However, due to their unique properties brought about by their small size, safety concerns have emerged as their physicochemical properties can lead to altered pharmacokinetics, with the potential to cross biological barriers. The intrinsic toxicity of some of the inorganic materials (i.e., heavy metals) and their ability to accumulate and persist in the human body has been a challenge to their translation. Indeed, there are some instances where NPs enable analyses and therapies that simply cannot be performed otherwise. So far, toxicity data generated by employing various models are conflicting and inconsistent. This chapter briefly describes factors affecting nanomaterial toxicity and organ *and* non-organ-directed nanotoxicity, including the behavior of NPs in human health and disease.

4.2   Nanoparticles Versus Larger Particles

NP technology is a known field of research since the last century. Nanotechnology produced materials of various types at the nanoscale level. NPs are a wide class of materials that include particulate substances, which have one dimension less than 100 nm at least. Depending on the overall shape these materials can be 0D, 1D, 2D, or 3D. The importance of these materials was realized when researchers found that size can influence the physiochemical properties of a substance, for example, the optical properties. The NPs show characteristic colors and properties with variation in size and shape, which can be utilized in bioimaging applications. Two principal factors cause the properties of nanomaterials to differ significantly from other materials: increased relative surface area and quantum effects. These factors can change or enhance properties such as reactivity, strength, and electrical characteristics (see also Chap. 2). As a particle decreases in size, a greater proportion of atoms are found at the surface compared to those inside. For example, a particle of size 30 nm has 5% of its atoms on its surface, at 10 nm 20% of its atoms, and at 3 nm 50% of its atoms. Thus, nanoparticles have a much greater surface area per unit mass compared with larger particles. As growth and catalytic chemical reactions occur at surfaces, this means that a given mass of material in the nanoparticulate form will be much more reactive than the same mass of material made up of larger particles.

Table 4.1 summarizes some of the differences between NPs (<100 nm) and larger particles (>500 nm) that impact their biological/toxicological effects when taken up via the respiratory tract. There is no biologically plausible reason for a strict borderline of 100 nm that separates NPs from larger particles. For example, even 240 nm polystyrene particles can act like NPs and translocate across the alveolo-capillary barrier in the lung when they are coated with phospholipids.

Table 4.1. Nanoparticles vs. larger particles: characteristics, biokinetics, and effects (respiratory tract as the portal of entry)

	Nanoparticles (<100 nm)	Larger particles (>500 nm)
General characteristics		
Ratio: number/surface area per volume	High	Low
Agglomeration in air, liquids	Likely (dependent on medium: surface)	Less likely
Deposition in the respiratory tract	Diffusion: throughout resp. tract	Sedimentation, impaction, interception; throughout resp. tract
Protein/lipid adsorption in vitro	Yes, important for biokinetics	Less effective
Translocation to secondary target organs		
Clearance	Yes	Generally not (to liver under "overload")
Mucociliary	Probably yes	Efficient
Alv. Macrophages	Poor	Efficient
Epithelial cells	Yes	Mainly under overload
Lymphatic circulation	Yes	Under overload
Blood circulation	Yes	Under overload
Sensory neurons (uptake + transport)	Yes	No
Protein/lipid adsorption in vivo	Yes	Some
Cell entry/uptake	Yes (caveolae; clathrin; lipid raft; diffusion)	Primarily phagocytic cells
Mitochondria	Yes	No
Nucleus	Yes (<40 nm)	No
Direct effects (caveat: chemistry and dose!)		
At secondary target organs	Yes	No
At portal of entry (resp. tract)	Yes	Yes
Inflammation	Yes	Yes
Oxidative stress	Yes	Yes
Activation of signaling pathways	Yes	Yes
Primary genotoxicity	Some	No
Carcinogenicity	Yes	Yes

4.3   Nanoparticles for Medical Use

NPs are broadly divided into various categories depending on their morphology, size, and chemical properties. Based on physical and chemical characteristics, some of the well-known classes of NPs that are commonly used in medical practice are summarized below. NPs have drawn increasing interest from every branch of medicine for their ability to deliver drugs in the optimum dosage range often resulting in increased therapeutic efficiency of the drugs, weakened side effects, and improved patient compliance.

4.3.1   Metallic Substances

Metal NPs are purely made of metal precursors. Due to well-known localized surface plasmon resonance (LSPR) characteristics, these NPs possess unique optoelectrical properties. NPs of the alkali and noble metals, that is, Copper (Cu), Silver (Ag), and Gold (Au), have a broad absorption band in the visible zone of the electromagnetic solar spectrum. Due to their advanced optical properties, metal NPs find applications in many research areas including biomedical sciences.

Gold NPs are considered to be relatively safe, as their core is inert and non-toxic. However, there are some other reports suggesting that cytotoxicity associated with gold NPs depends on the dose, side chain (cationic), and the stabilizers used. Cytotoxicity of gold NPs is dependent on the type of toxicity assay, cell line, and physical/chemical properties. The variation in toxicity with respect to different cell lines has been observed in human lung and liver cancer cell line.

Aluminum-based (Al) NPs contribute 20% to all nano-sized chemicals. According to the Global Market for Aluminum Oxide NPs Organization, Al-based NPs are being used in many areas, such as fuel cells, polymers, paints, coatings, textiles, biomaterials, etc. (http://www.futuremarketsinc.com). Al_2O_3 NPs disturb the cell viability, alter mitochondrial function, increase oxidative stress, and also alter tight junction protein expression of the blood-brain barrier (BBB). Besides, these NPs possess dose-dependent genotoxic properties and may have other toxic health effects on humans.

Zinc oxide (ZnO) NPs have many applications and are being used in paints, wave filters, UV detectors, gas sensors, sunscreens, and many personal care products. On the basis of increased use in many areas, human exposure to ZnO NPs is imminent. Cytotoxicity, cell membrane damage, increased oxidative stress, change in cell morphology, DNA damage, alteration in mitochondrial activity in human hepatocytes, and embryonic kidney cells have been observed. Besides cytotoxicity, the genotoxic potential of ZnO NPs has been reported.

Titanium oxide (TiO_2) is chemically an inert compound, but studies have shown that NPs of titanium dioxide possess some toxic health effects in experimental animals, including DNA damage as well as genotoxicity and lung inflammation. TiO_2 NPs induce oxidative stress and form DNA adducts. Besides genotoxicity, TiO_2

NPs possess toxic effects on immune function, liver, kidney, spleen, myocardium, glucose, and lipids homeostasis in experimental animals.

Iron oxide (Fe_2O_3) NPs have been used in biomedical, drug delivery, and diagnostic fields. These NPs bioaccumulate in the liver and other reticuloendothelial system organs. In vivo studies have shown that after entering the cells, Fe_2O_3 NPs remain in cell organelles (endosomes/lysosomes), release into the cytoplasm after decomposing, and contribute to cellular iron poll. Magnetic Fe_2O_3 NPs have been observed to accumulate in the liver, spleen, lungs, and brain after inhalation. Evidence show that these NPs exert their toxic effect in the form of cell lysis, inflammation, and disturbing blood coagulation system. Also, reduced cell viability has been reported as the most common toxic effect of Fe_2O_3. The toxic effects of Fe_2O_3 NPs are due to excessive production of ROS. These generated ROS further elicit DNA damage and lipid peroxidation.

4.3.2 Non-metallic Substances

Non-metallic substances such as carbon nanotubes (CNTs) and fullerenes are the most attractive and are widely used NMs. Fullerenes contain nanomaterials that are made of globular hollow cage such as allotropic forms of carbon. These materials have created noteworthy commercial interest due to their electrical conductivity, high strength, structure, electron affinity, and versatility. These materials possess arranged pentagonal and hexagonal carbon units. CNTs are elongated, tubular structures, 1–2 nm in diameter. These structurally resemble graphite sheets rolling upon themselves. The rolled sheets can be single, double, or many walls, and therefore they are named single-walled (SWNTs), double-walled (DWNTs), or multi-walled carbon nanotubes (MWNTs), respectively.

Due to their unique physical, chemical, and mechanical characteristics, these materials are not only used in pristine form but also in nanocomposites for many commercial applications such as fillers, efficient gas adsorbents for environmental remediation, and as a support medium for different inorganic and organic catalysts. Carbon-based NMs possess size-dependent carcinogenic effects similar to asbestos after injecting into the peritoneal cavity. Exposure to certain fibers shows some resemblance between CNTs and some other high aspect ratio (long and thin) NPs. When nanotubes, possibly of any chemical composition, have similar characteristics as some types of hazardous asbestos, similar inflammatory reactions can be induced by the CNTs. The main characteristics required for this to occur are long and thin fibrous forms (length $> 20\mu m$), rigidity, and non-degradability (biopersistence). Whether such nanotubes would pose a risk for humans is unknown. However, CNTs seem to conform to the same paradigm as some forms of asbestos, glass fibers, etc., that any long, thin biopersistent fiber poses a potential mesothelioma hazard. This means that other high aspect ratio NPs such as nanowires or nanorods are likely to have the same hazard if they satisfy the criteria of length and biopersistence.

4.3.3 Polymeric Materials

Biodegradable or polymeric NPs have the potential to be used in targeted drug delivery in cancer chemotherapy. These NPs are also employed in the encapsulation of various molecules to develop nanomedicine providing sustained release and good biocompatibility with cells and tissues. In addition, they have the potential to be successfully used in the encapsulation of peptides, nucleic acids, and proteins. They are also considered non-toxic, non-immunologic, and non-inflammatory and do not activate neutrophils. Poly-(DL-lactide-co-glycolide) has been used very successfully as a nanosystem for the targeted delivery of drugs and other molecules with the least toxicity, as it undergoes hydrolysis and produces biocompatible metabolites, lactic acid, and glycolic acid. However, surface coating induces the toxicity of polymeric NPs toward human-like macrophages.

4.3.4 Ceramic Materials

Ceramic (often in the form of mesoporous silica) NPs are inorganic nonmetallic solids, synthesized via heat and successive cooling. They can be found in amorphous, polycrystalline, dense, porous, or hollow forms. Therefore, these NPs are getting great attention from researchers due to their use in applications such as catalysis, photocatalysis, photodegradation of dyes, and imaging applications. The use of Silica (Si) NPs has many advantages in drug delivery systems. These NPs can be easily functionalized as drug carriers. Si NPs can lead to the generation of ROS; subsequent oxidative stress; induction of inflammatory biomarkers such as IL-1, IL-6, IL-8, and TNF-α (tumor necrosis factor); mitochondrial damage; and hepatotoxic effects.

4.3.5 Semiconductor Materials

Semiconductor NPs are known to possess a wealth of quantum phenomena and show unique size-dependent material properties. These materials possess properties between metals and nonmetals and therefore they found various applications in different fields. Semiconductor NPs possess wide bandgaps and therefore showed significant alteration in their properties with bandgap tuning. They are very important materials in photocatalysis, photo optics, and electronic devices. For example, a variety of semiconductor NPs are found exceptionally efficient in water-splitting applications, due to their suitable bandgap and bandedge positions. Copper oxide (CuO) NPs are used in semiconductors, anti-microbial reagents, heat transfer fluids, and intrauterine contraceptive devices. Experimentally, Cu NMs have been documented to possess toxic effects on the liver, kidney, and spleen in experimental animals. After oral administration and interaction with gastric juice, highly reactive ionic Cu is formed, which is then accumulated in the kidney of exposed animals. In one in vitro study, CuO NPs (50 nm) have been reported as being genotoxic and cytotoxic along with disturbing cell membrane integrity and inducing oxidative stress.

4.4 Applications in Biomedicine

Nano-sized inorganic particles of either simple or complex nature display unique physical and chemical properties and represent an increasingly important material in the development of novel nano devices which can be used in numerous physical, biological, biomedical, and pharmaceutical applications. Historically, Ag has long been known as an antibacterial substance. Ag NPs are being used increasingly in wound dressings, catheters, and various household products due to their antimicrobial activity. Antimicrobial agents are extremely vital in textile, medicine, water disinfection, and food packaging. Therefore, the antimicrobial characteristics of inorganic NPs add more potency to this important aspect, as compared to organic compounds, which are relatively toxic to biological systems. These NPs are functionalized with various groups to overcome the microbial species selectively. TiO_2, ZnO, Bismuth Vanadate ($BiVO_4$), and Cu- and Nickel (Ni)-based NPs have been utilized for this purpose due to their suitable antibacterial efficacies. Other applications include MRI contrast enhancement, tissue repair, immunoassay, fluorescent biological labels, biodetection of pathogens, detection of proteins, probing of DNA structure, tumor destruction via heating (hyperthermia), separation and purification of biological molecules and cells, phagokinetic studies, drug delivery, and cell separation. The detection of analytes in tissue sections can be accomplished through antigen-antibody interactions using antibodies labeled with fluorescent dyes, enzymes, radioactive compounds, or colloidal Au. The controlled release of pharmacologically active drugs to the precise action site at the therapeutically optimum degree and dose regimen has been a major goal in designing such devices. Semiconductor and metallic NPs have immense potential for cancer diagnosis and therapy on account of their surface plasmon resonance (SPR) enhanced light scattering and absorption. In addition, the antineoplastic effect and immune system therapy of NPs is also effectively employed to inhibit the tumor growth (see also Chap. 2).

4.5 Factors Influencing Nanotoxicity

Physicochemical properties such as particle size, distributions, geometry, and dimensions, surface charge, steric interactions, and agglomeration, surface area and reactivity, surface chemistry, surface morphology, surface impurities, bio persistence, and chemical transformation of engineered NMs are known to affect biological/toxicological effects (see also Chap. 2).

In addition, some other factors that influence toxicity include dose and exposure time, particle size, and their crystal structure. These factors result in different cellular uptake and affect oxidative mechanisms and subcellular localization. Likewise, pre-exposure and lower NP concentration can stimulate the cellular phagocytic activity (pre-exposure results in the adaptability of the human body against NPs to some degree).

4.6   Entry Sites, Fate, and Cellular Membrane Interactions

The human body has different entry portals for drug administration, such as the lungs via inhalation, the gastrointestinal tract via digestion, the blood vessels via intravenous (iv) injection, and the skin via transdermal passing. There is a range of potential human exposures that include occupational, due to the direct manufacturing process or a byproduct from an industrial or office environment, as well as incidental, from contaminated outdoor air and other byproduct emissions. For medical application, injection (intravenous, intramuscular, subcutaneous, and other) will also be an important entry route into the organism. Additives of NP/NM to food (TiO2; SiO2) and potential contamination of food from nano-enabled packaging materials result in exposure via the GI tract. Based on available data, translocation of nanomaterials in vivo across GI-tract epithelial cells seems to be limited. Skin exposure via cosmetics and skin-care products occurs, although penetration of healthy skin by NP may not occur. After entry, the fate of NPs, cellular uptake pathways, and cellular membrane interactions have been briefly summarized in Chap. 3.

4.7   Health Exposure Concerns of Nanoparticles

NPs have diameters between 1 and 100 nm that may be in the gas, liquid, powder, or embedded in the matrix. The precise meaning can be determined by the shape as well as the diameter of the NPs measured. The morphologies might differ extensively at the nanoscale. For instance, fullerenes are spherical whereas single-wall-carbon nanotubes (SWCNTs) are cylindrical. Potential exposure may arise from the environment during their production, development, use, or discarding or from therapeutic or medical use.

Ultrafine particles (UFPs) are not purposefully manufactured nor are they necessarily of a constant composition or size. They have been used to define aerosol and airborne particles less than 100 nm in diameter (so they are nano-sized), whereas NPs are manufactured through controlled engineering processes. The majority of preclinical and clinical studies on UFPs have been conducted with diesel exhaust and diesel exhaust particles (DEPs), an especially rich source of UFPs (Fig. 4.1).

Engineered NPs are nanoscale particles which are products of processes involving combustion and vaporization which are designed with very specific physical and chemical properties that make them very attractive for commercial development.

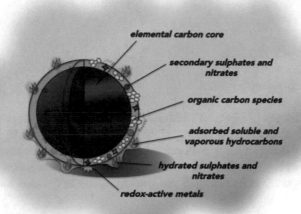

Fig. 4.1 Schematic providing an example of the complex composition of UFPs [e.g., urban particulate matter (PM) or particles in vehicle exhaust], which in urban air often have a carbon core coated with a diverse range of chemical species including reactive transition metals and organic hydrocarbons. Detail is not to scale
https://www.ncbi.nlm.nih.gov/pmc/articles/PMC5933410/bin/EHP424_f2.jpg

They have found applications in cosmetics, clothes electronics, biomedicine, aerospace, and computer industry. Due to their small size and large surface area, engineered NPs may have a high rate of pulmonary deposition and translocation ability to travel from the lung to systemic sites, penetrate dermal barriers, and a high inflammatory potency per mass. (Aerosol is a suspension of fine solid particles or liquid droplets in a gas. It includes smoke, air pollutants, and perfume spray. A nano-aerosol, therefore, comprises of NPs suspended in a gas, and may be present as discrete particles, or as clusters of NPs. However, there is not much literature on the health implication of gaseous NPs, and studies are needed to identify mechanisms by which nano-aerosols induce cellular damage and generate oxidative stress.)

In order to quantify exposure and risk, both in vivo and in vitro studies of various NPs and UFP species are currently being done using a variety of animal models including mouse, rat, and fish. These studies aim to establish toxicological profiles necessary for risk assessment, risk management, and potential regulation and legislation.

4.8 Nanotoxicity

There is a range of potential human exposures that include occupational, due to the direct manufacturing process or a byproduct from an industrial or office environment, as well as incidental, from contaminated outdoor air, biomedical applications, and other byproduct emissions. Exposure to these NPs could lead to either adverse effects at the organ level or non-organ-directed toxicity such as genetic effects, carcinogenicity, or adverse effects on the immune system.

4.8.1   Organ Toxicity

4.8.1.1   Respiratory System

The main exposure to UFPs/NPs is through inhalation, and they generally enter the body through the lungs and subsequently translocate to essentially all organs. The respiratory system is a network of organs that deals with the process of respiration, that is, moving from the nose through the trachea to the bronchioles (Fig. 4.2). The system is responsible for taking in and sending out air in living animals.

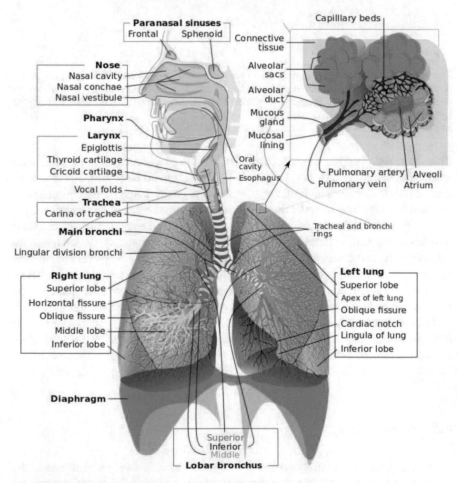

Fig. 4.2 The respiratory system (airway and lung anatomy)
https://www.ncbi.nlm.nih.gov/books/NBK470562/bin/845px-Respiratory_system_complete_en.svg.jpg

The lungs are part of the respiratory tract responsible for the exchange of gases. Inhalation is the most common route of exposure to NPs in the workplace. Once inhaled, these materials are carried by electrostatic force of the air from the upper to the lower respiratory tract. The particles are usually inhaled in the form of airborne NPs, systemic administration of drugs, chemicals, and other compounds to the lungs through direct cardiac output to the pulmonary arteries.

It has been shown experimentally that the biokinetics of NPs in the body differs depending on the portal of entry. The same NPs administered to the lung (inhalation or intratracheal instillation) or intravenously interact with different biological media and will receive different secondary coatings. Their entry into the blood circulation is at different dose and dose rates and into blood of different oxygenation states, all of which affect NP biodistribution to secondary target organs (Fig. 4.3). The figure demonstrates fundamental differences in NP transfer routes to blood and body organs when NPs are administered into the respiratory tract. Whereas dose and dose rates associated with NP administration by direct iv injection are obviously very high, both dose and entry rate of NPs from lung deposits into the blood compartment (arterial) are low. Thus, NPs delivered to the respiratory tract enter the blood compartment not only at different dose levels and different dose rates than directly iv injected NPs but also into blood of different oxygenation states (arterial versus venous).

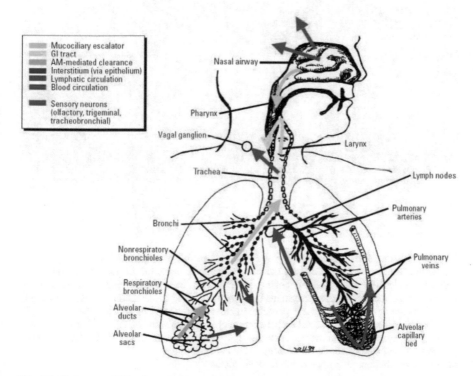

Fig. 4.3 Pathways of NPs through the respiratory system
https://www.ncbi.nlm.nih.gov/pmc/articles/PMC1257642/bin/ehp0113-000823f9.jpg

Immediately the NPs in the pulmonary sites translocate to blood circulation through the lymphatic pathways. Inhaled UFPs modify numerous aspects of cardiac function, reducing heart rate variability, a predictor of cardiovascular risk, and increasing the incidence, duration, and severity of arrhythmia. Furthermore, UFPs in urban air or diesel engine emissions exacerbate myocardial ischemia. Blood vessels finely regulate blood flow through changes in the tone of vascular smooth muscle, and UFPs generally alter the balance in favor of constriction. The resulting increased blood pressure and the reduced ability of the arteries to relax are usually detrimental. Vascular dysfunction can be caused by a loss of mediators such as nitric oxide released by the vascular endothelium, by increased sensitivity to vasoconstrictor factors, and by alterations in baroreceptor/neuroregulatory feedback. Blood components are also dysregulated, with UFPs tending to increase blood coagulability, encourage platelet activation, and reduce blood clot clearance. The cellular and

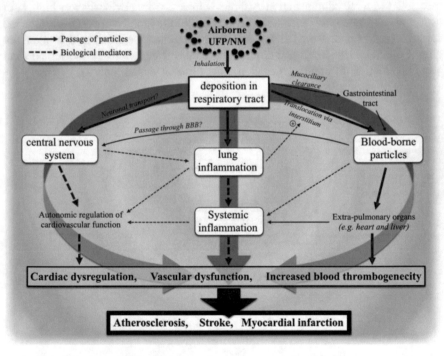

Fig. 4.4 Schematic illustrating some of the key mechanisms through which inhaled UFPs may influence secondary organs and systemic tissues, with emphasis on the means through which inhaled particles may cause cardiovascular events. Note that there are three main pathways linking the pulmonary and cardiovascular systems (*grey arrows*, left to right): autonomic regulation, passage of inflammatory mediators, and particle translocation. The arrows between these three pathways highlight the degree of interaction between mechanistic pathways and the challenges involved in the broad categorization of the wide-ranging biological actions of inhaled UFPs. Added to these pathways is the potential for desorbed components to exert effects
https://www.ncbi.nlm.nih.gov/pmc/articles/PMC5933410/bin/EHP424_f3.jpg

biochemical mechanisms underlying these effects are wide ranging, with oxidative stress and inflammation being key drivers (Fig. 4.4). In combination, these actions promote cardiovascular disease. Indeed, long-term exposure to UFPs in animal models has been shown to worsen atherosclerotic vessel disease.

Clearance from Respiratory Tract

Once NP/UFPs are deposited in the respiratory tract, there are several differences that separate NPs from larger particles; alveolar macrophages generally are attracted to deposited particles by chemotactic signals generated at the site of deposition. Nano-sized particles may be too small to generate such signals; studies in rodents with inhaled particles ranging from nano- to micro-size have shown that in vivo macrophage surveillance in terms of phagocytosis and clearance of nano-sized particles is rather poor so that deposited NP will interact with epithelial cells. Uptake into the pulmonary interstitium occurs, and there is evidence that some of the interstitialized NP reenter the airways, possibly the smaller conducting airways, from where they are cleared via the mucociliary escalator. Translocation into the interstitium and subsequently into blood and lymph circulation distinguishes nanoparticles from macroparticles.

Pathobiological Processes

Since NPs can cause severe lung diseases and impair lung function, it is of great importance to understand the mechanisms by which these diseases occurred. Four pathobiological processes considered most relevant to lung injury are as under:

(a) Oxidative stress.
(b) Inflammation
(c) Genotoxicity
(d) Fibrosis

These processes of the mechanism of nanotoxicity have been discussed in greater detail in Chap. 3.

4.8.1.2 Cardiovascular System

The cardiovascular events including myocardial infarction and heart failure are attributed to the exposure to combustion-derived NPs that incorporate reactive organic and transition metal components. After inhalation, the resulting respiratory inflammation induces systemic effects either directly by translocation from the lung or indirectly by yet unknown mediators. In view of these findings, a risk for NPs interaction with the cardiovascular system can be imagined. Mixed carbon NPs such as SWCNs, MWCNs, but not C60 fullerenes can stimulate platelet aggregation in vitro, and accelerate vascular thrombosis in a ferric chloride model of thrombosis in a specific rat model. However, modified fullerenes {Fullerenol, C60(OH)24} can facilitate adenosine diphosphate (ADP)-induced platelet aggregation in vitro, whereas carbon black cannot. There seems to be a certain degree of cardiotoxicity associated with IO NPs, but the systemic effect of these microscopic changes is yet to fully be determined by current research. Several NPs/NMs designed for drug delivery purposes, including alcohol/polysorbate NPs, gadolinium NPs, and nanostructured silica hydroxyethyl methacrylate biocomposites, do not have adverse effects on platelet function in vitro. In conclusion, the information on the possible hazard of NPs for cardiovascular effects is rather limited and needs further investigation.

4.8.1.3 Central Nervous System

The central nervous system (CNS) is complex with many nuances that are not yet elucidated within the system itself, especially the brain. The variability of toxicity effects on neural cells greatly depends upon the cell type and composition of the nanoparticles as has been seen in other organ systems as well.

In the central nervous system (CNS), the basic translocation pathways of inhaled NP are from the upper and lower respiratory tract, and through the blood-brain barrier (Fig. 4.5). The high diffusional deposition of the smallest NPs in the upper respiratory tract has significant biological/toxicological implications because NPs of this size range behave similarly to smell molecules in the inspired air that are directed at the olfactory mucosa. For efficient smell recognition, numerous neurons embedded in the olfactory mucosa are connected to the nasal lumen through their

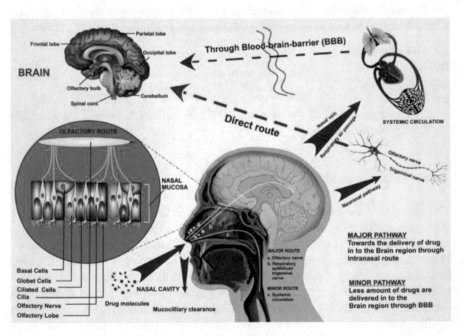

Fig. 4.5 Nanocarriers via nose-to-brain pathway for the central nervous system
https://media.springernature.com/lw685/springer-static/image/art%3A10.1007%2Fs11064-021-03488-7/MediaObjects/11064_2021_3488_Fig4_HTML.png

dendrites. The significance of sensory nerve structures lies in the potential that neuronal axons and dendrites can transport nanoparticles in retrograde and anterograde directions. Nanoparticles deposited in the nasal cavity can translocate with amazing velocity along this sensory neuronal pathway to the olfactory bulb. The nasopharyngeal airways and the tracheobronchial airways are supplied with sensory nerves as well, whereas the presence of sensory nerves in the alveolar region of the lung is less defined. These sensory nerves have either direct (olfactory and trigeminus nerve) or indirect (tracheobronchial, via vagus) connections to specific areas of the CNS. Thus, translocation of NPs depositing in the respiratory tract can occur along different routes. NPs depositing in the upper and lower respiratory tract may translocate directly to the blood compartment or via the nasal and lung lymphatic circulation. However, translocation rates for NPs into the blood circulation appear to be very low.

4.8.1.4 Gastrointestinal

The gastrointestinal tract (GI tract, GIT, digestive tract, digestion tract, alimentary canal) is the tract from the mouth to the anus. This system is a complex, multi-organ system including the pharynx, esophagus, stomach, small and large intestines, rectum, liver, pancreas, and gallbladder. The primary findings related to the GI toxicity of NPs depend on NP size and composition. Specifically, studies on Au NPs have demonstrated that small particles (5 nm) preferentially produced pathological changes in the liver, whereas medium and large particles (20 and 50 nm) tended to target the spleen. The toxic histopathological changes caused by the small Au NPs in the liver included steatosis, cytoplasmic degeneration, infiltration of inflammatory cells, Kupffer cells activation, and hemorrhage. Another prominent type of nanoparticles used in biomedical applications is IO NPs or CuO NPs. IO NPs tend to accumulate in the liver and other reticuloendothelial system (RES) organs. The iron products are recycled and then incorporated into pathways involved in hemoglobin, ferritin, and transferrin. However, toxicity studies have indicated dose-dependent toxicity. Comparative toxicology research among SiO_2, Ag NPs, and IO NPs indicates the Ag NPs can cause a greater degree of GI systemic toxicity as demonstrated by increases in serum alkaline phosphatase and calcium, lymphocytic infiltration in the liver which is not observed in SiO_2 and IO NPs.

4.8.1.5 Renal (Kidney)

Compared to the liver, the kidneys are less sensitive to NP, especially with Au NPs. Within the kidneys, the proximal tubule epithelial cells are the primary targets of nanotoxicity. However, Au NP toxicity is dose- and time-dependent. This indicates that the toxic potential of gold nanoparticles at a reduced concentration is quite minimal and displays some degree of reversibility as cell growth adapts and becomes resistant to change. Renal toxicity effects of IO NPs and Au NPs are almost similar except the distribution of iron in the kidneys is quite limited, with smaller particles being rapidly cleared in the urine. However, lower concentrations of IO NPs do not show any significant change in the architecture of the kidneys.

4.8.1.6 Reproductive System

The reproductive system includes the organs involved in producing offspring. In women, this system includes the ovaries, the fallopian tubes, the uterus, the cervix, and the vagina. In men, it includes the prostate, the testes, and the penis. Reproductive toxicity due to NPs requires further differentiation between their effects on male versus female reproductive systems. In general, NPs have been demonstrated to cross the blood-testicle and blood-placenta barrier, pressing the importance of addressing reproductive toxicity. Some studies have shown a decrease in sperm motility, albeit only at very high concentrations of Au NPs. In females, Au NPs and IO NPs show greater accumulation within the uterus. In addition, the negative impact on female sex hormones is largely seen with TiO2 NPs as they increase expression of the Cyp17a1 gene which in turn increases estradiol, apoptotic-related genes, inflammatory and immune responses, among other effects. Furthermore, some NPs have been shown to produce morphological changes in the follicles leading to a reduction in the mature oocytes present. Ovarian toxicity is observed with long-term TiO2 NPs, which cause a shapeless follicular antrum and irregular arrangement of cells, though the results are inconclusive and not very general. Likewise, inhaled cadmium oxide NPs impact fetal and neonatal development and growth.

4.8.2 Non-organ-Directed Toxicity

4.8.2.1 Carcinogenic and Genotoxic Effects

A genotoxic substance deleteriously impacts the genome of a cell either by direct or indirect damage to the cellular DNA including effects on the cellular pathways that monitor and protect genome integrity. The carcinogenic and genotoxic effects of conventional NPs are driven by the two following mechanisms:

(a) Direct genotoxicity
(b) Indirect (inflammatory processes-mediated) genotoxicity

   Primary genotoxicity is caused by direct binding of the NP with the DNA or component of the cell division machinery such as centromeres or microtubule spindle or intrinsic free radical production. NPs may cause genotoxicity through both mechanisms, that is, by direct interaction between CNT and DNA or through the induction of chronic inflammation leading to persistent oxidative stress. Long-term exposure renders the cell vulnerable to DNA aberrations that consequently lead to mutagenesis.

Gold NPs (Au-NPs) are one of the most common metallic NPs used in the biomedical field for diagnosis, treatment, and imaging due to their special characteristics, such as biocompatibility, controllable particle size and shape, ease of synthesis, controllable distribution, ease of surface functionalization, chemical stability, plasmonic properties, and tunable optical properties. Such inorganic NPs have the ability to cause damage to the genetic material because of their capacity to cross cell membranes. The toxic potential of NPs includes chromosomal fragmentation, DNA strand breakages, point mutations, oxidative DNA adducts, and alterations in gene expression profiles and consequently may initiate and promote mutagenesis and carcinogenesis. Genotoxicity mediated by the generation of excess ROS, referred to as secondary genotoxicity, has been reported. The detailed mechanism of action of NPs has already been discussed (see also Chap. 3).

4.8.2.2   Immune System

With the vigorous development of nanometer-sized materials, nanoproducts are becoming widely used in all aspects of life. The immune system can be perturbed at different levels, resulting in either its suppression or its overstimulation. In medicine, nanoparticles (NPs) are used as nanoscopic drug carriers and for nanoimaging technologies. Thus, substantial attention has been paid to the potential risks of NPs on the immune system. Only recently, attention has been directed toward the immune system toxicity of NPs/NMs. The toxicities that represent the most common safety concerns and reasons for NP failure in the preclinical stage include erythrocyte damage; thrombogenicity (platelet aggregation, plasma coagulation, *disseminated intravascular coagulation* [DIC], and leukocyte procoagulant activity [PCA]); cytokine-mediated inflammation and cytokine storming; pyrogenicity (mainly due to bacterial endotoxin contamination); and anaphylaxis and other complement activation-mediated reactions, as well as recognition and uptake by the cells of the mononuclear phagocyte system (MPS). Therefore, all new chemical and biological entities require adequate investigations into their interactions with the immune system before their use in industry, biology, and medicine. Although in recent years, our understanding of nanoparticle interaction with components of the immune system has improved, many questions still require more thorough investigation and deeper understanding. Further mechanistic studies investigating particle immunomodulatory effects (immunostimulatory and immunosuppression) are required to improve our understanding of the physicochemical parameters of nanoparticles that define their effects on the immune system.

The potential toxicity of nano medicine used for immune therapy has been discussed in Chap. 9.

4.9   Concluding Remarks

Although nanotechnology has made accurate, precise, and rapid diagnosis creating a major development in the field of medicine, yet one is exposed to NPs through different means. The human body has different entry portals for drug administration, such as the lungs via inhalation, the gastrointestinal tract via digestion, the blood vessels via intravenous injection, and the skin via transdermal passing. Their small size and shape helps NPs to travel through the bloodstream and ultimately reach the target organs and non-target organs in the human body. Nanotechnology aims for effective drug delivery, better ways of organ regeneration, and development of nano drugs. Techniques used in nanotechnology include delivery of drugs into targeted cells by using NP carriers. Unfortunately, direct or indirect exposure or excess use of nano medicine has led to toxicity due to NPs. NPs are usually more toxic to some cell subpopulations than others, and toxicity often varies with cell cycle. Cells exposed to NPs may undergo repairable oxidative stress and DNA damage or be induced into apoptosis. Exposure to NPs may cause the cells to alter their proliferation or differentiation or their cell–cell signaling with neighboring cells in a tissue. Because of quantum size effects and large surface area-to-volume ratio, nanomaterials have unique properties, compared with their larger counterparts. The toxicity of nanoparticles may not be seen with larger particles that consist of the same material or element. However, in view of the potential toxicity of NPs, there is a need for tighter regulations of nanotechnology products as the safety of these products is being questioned after nanomaterials showed high toxicity, particularly in inhalation studies.

Further Reading

Auría-Soro C, Nesma T, Juanes-Velasco P, Landeira-Viñuela A, et al. Interactions of NPs and biosystems: microenvironment of NPs and biomolecules in nanomedicine. Nanomaterials. 2014;2019(9):1365. https://doi.org/10.3390/nano9101365.

Clarence S, Yah S, Iyuke E, Geoffrey S. A review of nanoparticle toxicity and their routes of exposures. Iran J Pharm Sci. 2012;8(1):299–314.

Damasco JA, Ravi S, Perez JD, Hagaman DE, Melancon MP. Understanding nanoparticle toxicity to direct a safe-by-design approach in cancer nanomedicine. Nanomaterials. 2020;10(11):2186. https://doi.org/10.3390/nano10112186.

Furxhi I, Murphy F, Mullins M, Arvanitis A, Poland CA. Nanotoxicology data for in silico tools: a literature review. Nanotoxicology. 2020;14:612–37.

Gupta PK. Chapter 15: Toxic effects of nanoparticles. In: Toxicology: resource for self study questions. 2nd ed. Seattle: Kinder Direct Publications; 2020a.

Gupta PK. Chapter 14: Toxicology of nanomaterial particles. In: Problem solving questions in toxicology – a study guide for the board and other examinations. 1st ed. Cham: Springer Nature; 2020b.

Gupta PK. Toxic effects of nanoparticles. In: Brain storming questions in toxicology. 1st ed. Boca Raton: Taylor & Francis Group, LLC/CRC Press; 2020c. p. 297–300.

Gupta PK. Fundamentals of nanotoxicology. 1st ed. New York: Elsevier; 2022.

Jean-Marie E, Moudilou EN, Lapied E. Harmful effects of nanoparticles on animals. J Nanotechnol. 2015;2015:3–10. https://doi.org/10.1155/2015/861092.

Khan I, Saeed K, Khan I. Nanoparticles: properties, applications and toxicities. Review article. Arab J Chem. 2014;12(7):908–31.

Oberdörster G, Oberdörster E, Oberdörster J. Nanotoxicology: an emerging discipline evolving from studies of ultrafine particles. Environ Health Perspect. 2005;113(7):823–39. https://doi.org/10.1289/ehp.7339.

Simkó M, Nentwich M, Gazsó A, Fiedeler U. 2010. How nanoparticles enter the human body and their effects there. Nano Trust Dossiers, no 003en November 2010. https://epub.oeaw.ac.at/0xc1aa5576%200x0024c7a5.pdf

Stone V, Miller MR, Clift MJD, Elder A, et al. Nanomaterials versus ambient ultrafine particles: an opportunity to exchange toxicology knowledge. Environ Health Perspect. 2017;125(10):106002. https://doi.org/10.1289/EHP424. https://www.ncbi.nlm.nih.gov/pmc/articles/PMC5933410/#c231

Warheit DB, Oberdörster G, Kane AB, et al. Nanoparticle toxicology. In: Klaassen CD, editor. Casarett and Doull's toxicology: the basic science of poisons. 9th ed. New York: McGraw-Hill Education; 2019. p. 1381–430.

Yah CS, Geoffrey Simate SG, Iyuke ES. Nanoparticle toxicity and their routes of exposures. Pak J Pharm Sci. 2012;25(2):477–91.

Chapter 5
Nanotoxicity: Health and Safety Strategies

Abstract Commercializing various types of engineered NMs for a range of applications, including industrial, consumer, medical, and diagnostic, clearly presents a challenge for companies and regulators to ensure the development of safe and effective products for consumers. With regard to a general testing approach for human health hazard evaluation of NPs, a first step to determine potency may include a prioritization-related in vitro screening strategy to assess the possible reactivity, biomarkers of inflammation, and cellular uptake of NPs; however, this process should be validated using in vivo techniques. Of the possible hazards, inhalation exposure appears to present the most concern showing pulmonary effects such as inflammation, fibrosis, and carcinogenicity for some NMs. Skin contact and ingestion exposure, and dust explosion hazards, are also a concern. This chapter briefly describes nanotoxicity testing strategies and assessments of potential hazards associated with NMs.

Keywords Nanotechnology · Nanosafety · Exposures · Safety assessment · NPs · Nanoscale · Hazards · Risk assessment · Test procedures · NMs · Testing strategies · Nano biomedicine

5.1 Introduction

Nanotoxicology is a new area of study that deals with the toxicological profiles of nanomaterials (NMs). Compared with the larger counterparts, the quantum size effects and large surface area to volume ratio brings NMs their unique properties that may or may not be toxic to living things. It is therefore essential to assess the health and safety hazards including the potential toxicity of various types of NMs, as well as fire and dust explosion hazards. Because nanotechnology is a recent development, the health and safety effects of exposure to NMs, and what levels of exposure may be acceptable, are subjects of ongoing research. This chapter briefly

P K Gupta, *Nanotoxicology in Nanobiomedicine*,
https://doi.org/10.1007/978-3-031-24287-8_5

describes testing strategies and assessments of potential hazards associated with health risk assessments.

5.2 Sustainable Nanotechnology

There has been a surge of consumer products that incorporate nanoparticles (NPs), which are used to improve or impart new functionalities to the products based on their unique physicochemical properties. With such an increase in products containing NMs, there is a need to understand their potential impacts on the environment, ecosystem, animals, and human beings. This is often done using various biological models that are abundant in the different environmental compartments where the NMs may end up after use. In addition, we are familiar with the concept of "side effects": This is when something that is designed to be helpful ends up having some harm that goes along with it. It is well known that these drugs can also have some nasty side effects, yet people still choose to use them because the benefits of being cured outweigh the problems (and, of course, researchers continue to look for ways to reduce those negative effects). Nano medicine isn't the only area where this happens: There are many technologies that have lots of benefits to their use but can come with some potentially bad side effects. NPs are one example; they have many amazing uses for consumer products, but they can sometimes have harmful impacts on environmental organisms. The goal of sustainable nanotechnology is to try to find options for designing more environmentally friendly NPs so that those harmful impacts can be reduced.

5.3 Regulation

There is ongoing controversy on the implications of nanotechnology and a significant debate concerning whether nanotechnology or nanotechnology-based products merit special government regulation. This mainly relates to when to assess new substances prior to their release into the market, community, and environment. Regulatory bodies such as the United States Environmental Protection Agency (EPA) and the Food and Drug Administration (FDA) in the United States or the Health and Consumer Protection Directorate of the European Commission (EC) have started dealing with the potential risks posed by NPs. So far, neither engineered NPs nor the products and materials that contain them are subject to any special regulation regarding production, handling, or labeling. Based upon available data, it has been argued that current risk assessment methodologies are not suited to the hazards associated with NPs; in particular, existing toxicological and ecotoxicological methods are not up to the task; exposure evaluation (dose) needs to be expressed as quantity of NPs and/or surface area rather than simply mass; equipment for routine detecting and measuring NPs in air, water, or soil is inadequate;

and very little is known about the physiological responses to NPs. Regulatory bodies in the world, in particular the United States and the European Union (EU), have concluded that NPs form the potential for an entirely new risk and that it is necessary to carry out an extensive analysis of the risk. The challenge for regulators is whether a matrix can be developed which would identify NPs and more complex nano-formulations which are likely to have special toxicological properties or whether it is more reasonable for each particle or formulation to be tested separately. However, the use of nanotechnology in various fields such as consumer products, industrial applications, agriculture, biomedicine, and others is growing rapidly. Nano biomedicines are just beginning to enter drug regulatory processes, but within a few decades could comprise a dominant group within the class of innovative pharmaceuticals. Therefore, there is a need to develop comprehensive regulation of nanotechnology that will be vital to ensure the potential risks associated with use of nanomaterials during the research and commercial application of nanotechnology. Regulation may also be required to meet community expectations about the responsible development of nanotechnology, as well as to ensure that public interests are included in shaping the development of nanotechnology.

5.3.1   European Union REACH

REACH regulation aimed at ensuring the safe production, use, and import of substances entered into force in June 2007. Although there are no specific regulations of engineered NMs in the EU, REACH legislation requires an assessment of engineered NMs within the registration of the bulk form of a substance. In Europe, manufacturers and importers of carbon nano products, including carbon nano-tubes, will have to submit full health and safety data within a year or so in order to comply with REACH (Registration, Evaluation, Authorization, and Restriction of Chemicals).

The involvement of computational specialists in nano-safety research has become more prominent since REACH regulation promoted the use of in silico techniques, such as quantitative structure-activity relationship (QSAR) and read-across, for the purpose of risk assessment. However, the use of recently developed nano-QSAR models still requires support from classical laboratory methods to be accepted by the regulatory authorities and the end-users. Confidence in in-silico predictions can be only gained through the validation of computational models with "real-life" results. Other key issues that need to be considered in order to improve the regulatory acceptance of in silico models include:

- The uncertainties in the constructed models and the model's applicability domain should be clearly and transparently reported.
- In addition to the extracted knowledge and the derived computational models, the model builder should also attempt to provide some probabilistic reasoning to justify the results obtained.

- The modeler/reporter should use intelligible language, instead of complex technical terms, considering the background of end-users (i.e., experimentalists, industrial partners, or regulating authorities), who should have a clear understanding of the model and its applicability in order to avoid misuse of it.

5.3.2 United States

In the United States, rather than adopt a new nano-specific regulatory framework, the FDA convenes an "interest group" each quarter with representatives of FDA centers that have responsibility for the assessment and regulation of different substances and products. However, safety studies are required for each and every nano-science application.

5.3.3 International Law

Many regulatory systems around the world already assess new substances or products for safety on a case-by-case basis, before they are permitted on the market. These regulatory systems have been assessing the safety of nanometer-scale molecular arrangements for many years and many substances comprising nanometer-scale particles have been in use for decades, for example, carbon black, titanium dioxide, zinc oxide, bentonite, aluminum silicate, iron oxides, silicon dioxide, diatomaceous earth, kaolin, talc, montmorillonite, magnesium oxide, and copper sulfate. There is no international regulation of nano products or the underlying nanotechnology. Neither are there any internationally agreed definitions or terminology for nanotechnology, nor internationally agreed protocols for toxicity testing of NPs at present. Even no standardized protocols for evaluating the environmental impacts of NPs exist. However, general principles of toxicity as laid down by the Organization for Economic Cooperation and Development (OECD), EU, US FDA, or for that matter in any other developed country are being followed.

5.4 Toxicity to Direct a Safe-by-Design

As has been indicated previously, there are no specific regulations for the control and use of NMs. Therefore, a general approach for testing is being followed. It is well established that the consequent effects of NPs both at the cellular and systemic levels are highly dependent on their physicochemical properties (i.e., size, shape, composition, surface charges, and coating). For example, the surface charge has been critical in the non-specific binding and cellular uptake of the NPs. Au NPs coated with amphiphilic polymers of varying surface charges result in positively charged NPs having the highest uptake; however, this also leads to higher toxicity. Adsorption of serum proteins can be minimized by modifying the surface with *zwitterion* or neutral organic coatings while also yielding a small hydrodynamic size and high stability in biological media. The toxicity inherent to the core composition of inorganic NPs, especially those consisting of heavy metal atoms, and the leaching of ions from the dissolution of the NP core can be overcome by engineering the surface with a biocompatible coating. This strategy also regulates the high surface energy of the NPs while providing stability, bioavailability, and targeting. With these findings, interest in the development of ultrasmall sub-5 nm NPs, with judicious choices of surface coatings to improve the biocompatibility and pharmacokinetics of NPs, has been rapidly rising. However, NPs are highly heterogeneous, with very diverse combinations of chemical composition, core-shell structure, shape, and functionalization. This poses a challenge in the experimental assessment of the relationship between their physicochemical properties and their toxicological effects. Therefore, with regard to a testing approach for human health effects, the first step is to include in vitro screening assessment strategies to evaluate possible reactivity, biomarkers, inflammation, or cellular uptake indices. The in vitro screening assessment strategies is to be followed by "intelligent testing strategy" (ITS) designed to assess the risk of NMs, to be evaluated accurately, effectively, and efficiently, thereby obviating the need to test each and every NM type, on a case-by-case basis using physicochemical characterization, exposure identification, and hazard assessments.

5.4.1 Overview of In Vitro and Alternative Methods

The principle of the 3Rs—Replacement, Reduction, and Refinement—has become an increasing public and legal demand which ethically supports the replacement of animal use with more human-relevant alternatives that do not rely on in vivo testing. The specific properties of NMs such as smaller size but a larger surface area, and high catalytic reactivity and distinctive optical properties compared to their respective bulk entities, often disable a straightforward use of established in vitro approaches. However, the use of in vitro systems provides rapid and inexpensive results to predict the effects of NMs at the cellular level. Recent advances in the toxicity testing of NMs are beginning to shed light on the characteristics, uptake, and mechanisms of their toxicity in a variety of cell types.

New concepts for efficient, cheaper, and evidence-based testing strategies have been proposed, based on the use of human primary cells and cell lines. In addition, endpoints for health effects and in vitro tests of regulatory interest for conventional chemicals are contained in the OECD and its test guidelines documents (TG). These in vitro tests are rather narrow in their coverage of endpoints: They address genetic toxicity, dermal absorption, skin and eye irritation, endocrine disruption, and a few other selected endpoints. Skin penetration has not been a major concern for NMs while endocrine disruptor effects for NMs are also not currently a focus of research or regulatory concern. Rather, the most relevant in vitro protocols for NMs align with the current major routes of NM exposures. Besides dermal (NMs in cosmetic products) and oral (NMs in food products) exposures, the effects due to NM inhalation are currently considered to be the most relevant.

Cellular responses have been observed upon exposure to NMs and currently several hypotheses regarding how NMs induce adverse cellular effects exist: (i) via oxidative means (the oxidative stress paradigm) which then leads to pro-inflammatory effects, (ii) via the fiber paradigm, and (iii) through genotoxicity via NM dissolution, that is, release of potentially toxic ions and/or other constituents. The fiber paradigm has shown that multi-walled carbon nanotubes caused granulomas in the peritoneal cavity. This paradigm, however, can only be attributed to nanofibers, in particular to those with the specific characteristics of high rigidity and high aspect ratio NMs.

Other endpoints for NMs which can be examined in vitro include those that test for the biological fate of NMs at the cellular or multicellular levels, such as size exclusion criteria for given key cell types, and adverse effects, such as fibro-genicity at these levels of organization.

For assessing cell metabolic activity, the MTT assay (3-(4,5-dimethylthiazol-2-yl)-2,5-diphenyl tetrazolium bromide), which is a colorimetric assay, is used. NAD(P)H-dependent cellular oxidoreductase enzymes may, under defined conditions, reflect the number of viable cells present. These enzymes are capable of reducing the tetrazolium dye MTT to its insoluble formazan, which has a purple color. Other closely related tetrazolium dyes including XTT (2,3-bis(2-methoxy-4-nitro-5-sulfophenyl)-5-carboxanilide-2H-tetrazolium), MTS {3-(4,5-dimethylthiazol-2-yl)-5-(3-carboxymethoxyphenyl)-2-(4-sulfophenyl)-2H-tetrazolium)}, and the WSTs (water-soluble tetrazolium salts) are used in conjunction with the intermediate electron acceptor, 1-methoxy phenazine methosulfate (PMS). With WST-1, which is cell-impermeable, reduction occurs outside the cell via plasma membrane electron transport.

Tetrazolium dye assays can also be used to measure cytotoxicity (loss of viable cells) or cytostatic activity (shift from proliferation to quiescence) of potential medicinal agents and toxic materials. MTT assays are usually done in the dark since the MTT reagent is sensitive to light.

Current advances in machine learning—in silico, in vitro, and tissue model techniques—have placed predictive modeling of the in vivo response with an integrative alternative tiered toxicity testing strategy, a potential attainable long-term goal for nanotechnology risk assessment. Acellular assays, in vitro assays, and advances in "omics" coupled with computational modeling may provide a suite of biomarkers for high-throughput screening assays predictive of in vivo toxicity. Further model and method development in complex cell and tissue culture systems, including air–liquid interface culture, cellular co- and tri-cultures (3D), Zebrafish test, C. elegans model, biocorona and ADME systemic toxicology, organ-on-a-chip, and other alternative animal models, has placed in vitro screening techniques at the forefront of nanotoxicology testing for responsible nanotechnology development. The details of various in vitro and alternative tests are beyond the scope of this chapter.

These approaches may help to reduce and/or replace animal testing according to the 3R strategy. With all these goals, it is critical to use environmentally or occupationally relevant NM concentrations and to be able to relate these in vitro test concentrations to in vivo test exposures so that results can be correlated and used in a regulatory context.

5.4.2   Overview of In Vivo Evaluation

In order to evaluate the in vivo nanotoxicity of NMs, general principles of testing as laid down by each country are being followed. For example, the OECD, FDA, EU and REACH, and other countries of the world each have their guidelines for various toxicological test protocols to be followed for the safety evaluation of chemicals and pharmaceuticals. Some of them are briefly summarized as under:

Acute Oral Toxicity: For acute oral toxicity and lethal Dose 50 (LD50). Laboratory animals are given orally colloidal NMs in different doses and are observed for toxic symptoms and mortality.

Acute Eye Irritation and Corrosion Test: For acute eye irritation and corrosion test the colloidal suspension is placed in the conjunctival sac of one eye/animal and the other eye serves as a control and is treated with the same volume of distilled water. The eye reactions of conjunctivitis of cornea and chemosis are graded. The animals are maintained and observed daily over 14 days for various toxic symptoms.

Acute Dermal Toxicity Test: For acute dermal toxicity colloidal suspensions are applied to a shaved area of skin. The animals are maintained and observed daily over 14 days. Deaths and other skin reactions such as edema, erythema, ulcers, bloody scabs, discoloration and scars, and toxic signs (weight loss, water and food consumption, behavior) are recorded. At 1, 3, and 7 days after exposure, skin biopsies are performed for histopathological investigations. All animals are sacrificed after a 14-day observation period and skin is collected for routine histopathological examination.

Short- and Long-Term Studies: Short- and long-term studies have to be undertaken as per guidelines issued from time to time by regulatory agencies of each country. For the evaluation of chronic toxicity and carcinogenic potential of NMs, the OECD guidelines recommend the administration of substances for 12 and 24 months, respectively. After this period, the survival rate, the clinical toxicity signs, animal behavior, tumor incidences, and histopathological findings in the liver, spleen, kidneys, brain, ovary, and testis will be assessed. Likewise, protocols for other tests such as development and reproduction toxicity are also followed by different agencies.

Special Tests: To realize the optimal use of the NP platforms in the biological systems and to move forward with their clinical translation would require rational designs that are driven by how these physicochemical properties could impact their fate and effects in the body. Pharmacokinetics includes the absorption, distribution, metabolism, and excretion (ADME) of a drug or contrast agent. The four criteria influence the concentration of the substance and kinetics of the substance exposure to organs/tissues. For intravenous administration, the step of absorption is not involved because the substance is directly introduced to the systemic circulation (Fig. 5.1). For example, it has been shown that inorganic NPs with a hydrodynamic size of sub-5 nm can be cleared out of the body efficiently through renal clearance and still maintain efficient tumor targeting in comparison with those of higher sizes which accumulate rapidly in the organs of the RES depleting their availability for the intended target and increasing the risk of toxicity with prolonged exposures. This strategy also regulates the high surface energy of the NPs while providing stability, bioavailability, and targeting.

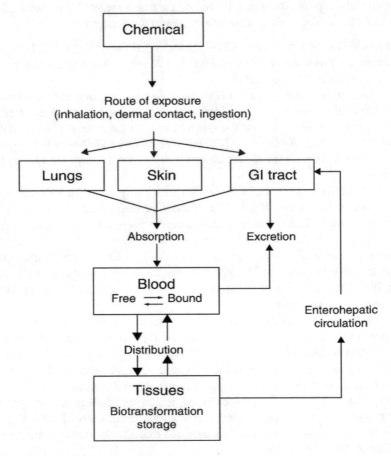

Fig. 5.1 Disposition and fate of toxicants
https://media.springernature.com/lw685/springer-static/image/
chp%3A10.1007%2F978-3-030-22250-5_2/MediaObjects/481219_1_En_2_Fig1_HTML.png

With these findings, interest in the development of ultrasmall sub-5 nm NPs, with judicious choices of surface coatings to improve the biocompatibility and pharmacokinetics of NPs, has been rapidly rising.

5.5   Toxicity Assessment

There are different testing (in vivo and in vitro), non-testing (in silico), and experimental models used for assessing the toxic effects of NPs. Cell-based tests and tissue engineering are in vitro approaches. In silico techniques include the application of prediction methods in nanotoxicity studies, such as molecular docking, quantitative structure–activity relationship (QSAR) assays, and molecular dynamics simulations (Fig. 5.2). Both in vitro and in silico approaches can be used to overcome ethical problems in medical research and nanotechnology. Given the complexity and heterogeneity of engineered NM classes, the implementation and future success of in silico models directly rely on the level of acceptance it gets from potential users and regulators and will be on a case-by-case basis. In order to increase the credibility of computational methods in nanotoxicology, it should be proven that the outcome of in silico models is (at least) as reliable as existing in vivo or validated in vitro tests. Usually in vitro and in vivo toxicological effects of NPs are conducted

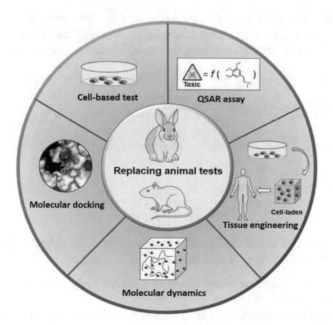

Fig. 5.2 An overview of the current alternatives to the use of animals, including cell-based tests, tissue engineering, and computer-based techniques for nanotoxicity assessment
https://www.mdpi.com/ijms/ijms-22-04216/article_deploy/html/images/ijms-22-04216-g001-550.jpg

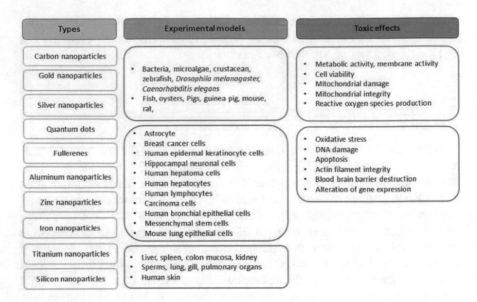

Fig. 5.3 Nanoparticle types, experimental models used for the studies, and toxic effects of NPs https://media.springernature.com/full/springer-static/image/art%3A10.1007%2Fs40089-017-0221-3/MediaObjects/40089_2017_221_Fig   2_HTML. gif?as=webp

on bacteria, microalgae, zebrafish, crustaceans, fish, rats, mice, pigs, guinea pigs, and human cell lines. NP types, experimental models used for the studies, and various toxic effects of NPs that are important to look into are summarized in Fig. 5.3.

5.5.1 Risk Assessment

The likelihood of an adverse effect to occur is assessed through four main steps:

(a) Hazard identification.
(b) Dose-response assessment.
(c) Exposure assessment.
(d) Risk characterization.

A complete risk assessment process consists of multiple steps with the aim of answering the following questions as accurately as possible:

- What harmful effects may be caused to the human body and the environment by a toxic substance?
- What quantitative correlations exist between the dose of a toxicant and the likelihood of adverse effect in an exposed population?
- What exposures are experienced or anticipated under different conditions?

- What are the severity, frequency, and probability of adverse effects in exposed populations?

The risk assessment process starts with the identification of the potential hazards associated with exposure to a hazardous chemical, for example, any human health or environmental problems that a chemical can cause. The second step is to determine the maximum tolerable or acceptable dose above which signs of adverse effects begin to occur (these are obtained from toxicology data, using in vivo experimental models). Generally, the higher the dose, the greater the likelihood of harmful effects to occur. For most substances, the safe levels are determined based on a threshold dose below which exposure poses no risks (the exceptions being carcinogens and mutagens where a dose that would result in an acceptable risk). The third step (i.e., exposure assessment) focuses on the identification of the exposed population and the determination of the exposure routes, amount, duration, and pattern. In the risk characterization phase, the data obtained from the previous steps of risk assessment are integrated to determine the probability of adverse human health and/ or environmental effects occurring as a result of exposure to a hazardous substance. It is also important to include uncertainty associated with the risk estimates in this phase.

Assessing the risk will deliver facts for a reasonable risk management. For NMs, chemical risk assessment is very often relevant. Here, two points need to be addressed. First, it is mandatory to identify and examine a potential hazard of the material under controlled conditions, for example, in the lab. The second mandatory point is to measure and quantify exposure, that is, the amount of material which comes into contact with humans, animals, or the environment, respectively. On the basis of these first two steps a specific risk can be characterized and a decision has to be made about the height of the risk for each exposure scenario (traffic light scheme).

Recently, for hazard and risk assessment, it has been suggested as under:

To identify, assess, and classify potential risks (considering cost and time), there is an urgent need for non-testing approaches in NP hazard risk assessment. One strategy that is being suggested for investigating the toxicological properties of a variety of NPs is using computational tools that decode how nano-specific features relate to toxicity and enable its prediction and is summarized in five domains:

- I. Data set formation.
- II. Data pre-processing.
- III. Model implementation.
- IV. Validation.
- V. Applicability.

A general roadmap for implementing a model in the field of nanotoxicology is as under:

1. Data Collection: Literature, experimental, databases, etc.
2. NP information: Types of NPs (metal, metal oxides, carbon-based, polymers, dendrimers, etc.); nano features (intrinsic, extrinsic, int/ext); and descriptors generation (DRAGON, MoE, MOPAC, etc.)

3. Study design information: System (in vitro, in vivo, combination, etc.); species (rodent, bacteria, human, etc.); tissue (lung, liver, or any other tissue); exposure conditions (duration, exposure dose, route, etc.); and in vitro tests (cell type, cell line, toxicology assay, etc.)
4. Endpoints: Cytotoxicity, ecotoxicity, or combined.

These results confirm the importance of the NP core material when considering the design of nanomedicines.

Computational nanotoxicology and algorithm-based approaches to predict the safety and efficacy of these NPs are emerging. With the ongoing development and validation of computational tools, accurate in silico predictions with regards to the safety and biological fate of the NP design can be achieved. Integration of computational modeling in the design stage that is highly useful in their advancement and success in the clinic is beyond the scope of this book.

5.5.2 Risk Management

Risk management of NMs is an integral part of risk analysis and is directly based on risk assessment. For NMs, usually, a chemical risk assessment is chosen. After a thorough risk assessment, it will arrive at a point where a decision must be taken. In Europe, the decision whether the assessed risk of a NM is seen to be unacceptable, acceptable under certain safety precautions, or completely acceptable is based on National or European Legislation. In fact, this has been done by bringing into force the European chemical legislation REACH in 2007. It controls the production and use of chemical substances, and their potential impacts on both human health and the environment. This regulation also includes NMs. It is managed and administrated by the European Chemicals Agency (ECHA).

In practice, risk management at the workplace, for instance, results in different measures, sometimes related to the production machine or process (special housing, closed system, air exhaust) and sometimes regarding the workers (breathing mask, gloves, helmet, etc.). In any case, the target is to reduce the exposure down below the threshold limits of the hazardous material. The same is true for consumers. In case of possible risk, there has to be a respective labeling of the product (e.g., ethanol as flammable, pesticides as toxic, etc.), and the consumer has to handle it with care.

5.5.3 Risk Communication

Once a risk of a new (nano) material is scientifically identified, it is important to transfer or "translate" the information to the decision-making parties and to the public. This needs to be done in a responsible way so that no misleading information is spread. This can be provided through a process called "risk communication."

5.6 Challenges and Future Strategies

During the past two decades, several research strategies and challenges have been proposed to facilitate and direct "verification," that is, the safe handling of and exposure to individual forms of NPs and nanotechnology in general. A major problem is that all of the different NP types cannot be effectively evaluated for safety and environmental effects in a timely manner owing to: (a) the vast numbers of different NP types; (b) the numerous variations within specific NP types (e.g., there are many different nanoscale forms of carbon: carbon black particles, fullerenes, single-walled carbon nanotubes, multi-walled carbon nanotubes [MWCNTs], carbon nanofibers [CNFs], etc.); (c) the overwhelming expense required to adequately test each individual NP type; (d) the inadequate time that would be required to test each of the NP types (e.g., acute tests may be studied for a short period of time, perhaps 24 h, but these types of studies do not represent longer-term exposures to humans or in the environment, i.e., subchronic or chronic lifetime studies, which in rats can be 2 years); and (e) the vast expense can easily cost heavy amount for a single study.

Exposure assessments of NMs are still underway and far from being complete due to the lack of available exposure measurement methods. Reliable routine airborne exposure measurement methods are not available, although exposure measurements of NMs at the workplace and/or in the environment are urgently needed. Moreover, the biological and environmental distribution pathways of NMs are largely unexplored but the situation improves with new publications.

Whether there is a potential risk of NMs cannot be answered in one simple statement. To give an answer, the varying hazard potentials of the different NMs and the variance in possible exposure scenarios have to be considered for a specific situation. Currently, risk assessment of manufactured NMs is done on a case-by-case basis. But testing of all existing NM variants is not feasible. One solution to overcome this problem is the grouping of NMs.

A few of the challenges are listed below:

- Advancement of equipment over the next few years to estimate or determine aerosolized and water exposures to engineered NMs.
- Techniques and approaches to assess the hazards of engineered NMs.
- Developing predictive models to gauge the potential hazardous effects of NM exposures on environmental and human health systems (environmental, health, and safety [EHS]).
- Developing methods for estimating the EHS impacts of NM exposures over a lifespan.
- Facilitating methodologies to strategically assess programs that could be utilized to implement relevant risk-focused research.

Keeping in view the various challenges, the following specific organizations have made a significant contribution in research and development for the safe use of nanotechnology.

- The US-based National Academy of Sciences, which published a comprehensive research plan for nanosafety research.
- The EU, which sponsored numerous research-based consortia.
- The OECD projects, which developed protocols to standardize and validate test methods for conducting toxicity studies on NMs.

5.7 Conclusion

Engineered NMs are increasingly produced and utilized in a wide variety of products, and the uncertainties regarding their potential adverse health effects, which are associated with their unique properties, make risk assessment challenging. Evidence of adverse health effects have been demonstrated in in vitro studies, in vivo studies, and in epidemiological studies. In many cases, however, the interpretation of available data is complicated by the use of very high doses to elicit cell responses and adverse effects; a lack of data relevant to chronic, low-dose exposures; complexities associated with NM characterization; and the dynamic nature of NMs in the environment and in biological systems. The rate at which new NMs with unique properties are developed is increasing and often outpaces the ability to characterize the potential adverse health risks associated with these materials. NPs can be redesigned to reduce their interactions with (and potential toxicity for) organisms in the environment. There is still a lot of work to do to understand the different ways NPs can potentially be toxic to organisms, and there are studies underway to understand these mechanisms further. However, we also have enough knowledge that can now be applied to investigate the various methods by which these side effects can be mitigated, which is a newer area to be explored in the field of NPs. It is important to remember that NPs have many great benefits for different consumer products and medical technologies, but by redesigning them to improve their environmental impact, we can work to make NP use more sustainable in the long run.

Further Reading

Bahadar H, Maqbool F, Niaz K, Abdollahi M. Toxicity of nanoparticles and an overview of current experimental models. Iran Biomed J. 2016;20(1):1–11. https://doi.org/10.7508/ibj.2016.01.001. https://www.ncbi.nlm.nih.gov/pmc/articles/PMC4689276/

Buchman JT, Hudson-Smith NV, Landy KM, Haynes CL. Understanding nanoparticle toxicity mechanisms to inform redesign strategies to reduce environmental impact. Acc Chem Res. 2019;52:1632–42. https://doi.org/10.1021/acs.accounts.9b00053.

Damasco JA, Ravi S, Perez JD, Hagaman DE, Melancon MP. Understanding nanoparticle toxicity to direct a safe-by-design approach in cancer nanomedicine – review. Nanomaterials. 2020; 10:2186, pp. 1–41. https://doi.org/10.3390/nano10112186.

European Parliament and Council. Regulation (EC). 2006. European Parliament and of the Council of 18 December 2006. No 1907/2006, concerning the Registration, Evaluation, Authorization and Restriction of Chemicals (REACH), establishing a European Chemicals Agency, amending Directive 1999/45/EC and repealing Council Regulation (EEC) No 793/93 and Commission Regulation (EC) No 1488/94 as well as Council Directive 76/769/EEC and Commission Directives 91/155/EEC, 93/67/EEC, 93/105/EC and 2000/21/EC. OJ EU 2006; L396:1.

Gupta PK. Chapter 15: Toxic effects of nanoparticles. In: Toxicology: resource for self study questions. 2nd ed. Seattle: Kinder Direct Publications; 2020a.

Gupta PK. Chapter 14: Toxicology of nanoparticles. In: Problem solving questions in toxicology – a study guide for the board and other examinations. 1st ed. Cham: Springer Nature; 2020b.

Gupta PK. Toxic effects of nanoparticles. In: Brain storming questions in toxicology. 1st ed. Boca Raton: Taylor & Francis Group, LLC, CRC Press; 2020c. p. 297–300.

Gupta PK. Fundamentals of nanotoxicology. 1st ed. New York: Elsevier; 2022.

OECD. Guideline for testing of chemicals. Paris: OECD; 2012.

Oksel C, Hunt N, Wilkins T, Wang XZ. Risk management of NMs: guidelines for the manufacture and use of nanomaterials. The Reach Centre; 2017, Jan 1. Version 1. http://www.sun-fp7.eu/wp-content/uploads/2017/01/SUN-risk-management-guidelines.pdf

Schrand AM, Dai L, Schlager JJ, Hussain SM. Toxicity testing of Nanomaterials. 2012. PMID: 22437813. https://doi.org/10.1007/978-1-4614-3055-1_5. pubmed.ncbi.nlm.nih.gov › 22437813

Simona C, Adriana F. In vivo assessment of nanomaterials toxicity. Open access peer-reviewed chapter; 2015. https://www.intechopen.com/books/Nanomaterials-toxicity-and-risk-assessment/in-vivo-assessment-of-NMs-toxicity. https://doi.org/10.5772/60707.

Warheit DB. Hazard and risk assessment strategies for nanoparticle exposures: how far have we come in the past 10 years? Version 1. F1000 Res. , 2018;7:376. Published online 2018 Mar 26. https://doi.org/10.12688/f1000research.12691.1. https://www.ncbi.nlm.nih.gov/pmc/articles/PMC5871814/

Warheit DB, Borm P, Hennes C, Lademann J. Testing strategies to establish the safety of nanomaterials: conclusions of an ECETOC workshop. Inhal Toxicol. 2007;19(8):631–43. https://doi.org/10.1080/08958370701353080.

Warheit DB, Oberdörster G, Kane AB, et al. Nanoparticle toxicology. In: Klaassen CD, editor. Casarett and Doull's toxicology: the basic science of poisons. 9th ed. New York: McGraw-Hill Education; 2019. p. 1381–430.

Chapter 6
Nano-Based Drug Delivery Systems

Abstract Nanoparticles are important for refining drug delivery. There has been a considerable increase in the number of new biotechnology-based therapeutics. In addition to being vehicles for drug delivery, NPs can be used as pharmaceuticals as well as diagnostics. Most of these are proteins and peptides, and their delivery present special challenges. Cell and gene therapies are sophisticated methods of delivery of therapeutics. Recently, advances in targeted drug delivery have occurred in therapy and the methods in which NCs can deliver drugs include: passive targeting, active targeting, vascular targeting (endothelial cells), targeting the mildly acidic tumor microenvironment (ph specificity), temperature specificity, and nuclear targeting. However, drug delivery to the brain across the blood–brain barrier presents many challenges. Therefore, refinements in drug delivery will facilitate the development of personalized medicine. This chapter highlights an overview of the current NC drug delivery systems (DDSs) starting with their properties, various routes of drug administration, various drug formulations, as well as devices used for drug delivery and targeted drug delivery. In this perspective, chemical, physical, and biological engineering principles and strategies for constructing DNA-assisted NCs or other suitable techniques that can be precisely reconfigured by external and internal stimuli to drive the release of a loaded drug in a target region with appropriate microenvironments have been discussed. Considering the potential applications of NPs in many fields and the growing apprehensions of the FDA about the toxic potential of nano products, it is the need of the hour to look for new internationally agreed free of bias toxicological models by focusing more on in vivo studies.

Keywords Controlled release · Drug delivery devices · Gene therapy · NPs · Medicine protein/peptide delivery · Routes of drug administration · Targeted drug delivery · Hydrogel · Micelle · Nanomaterials · Nanocarriers · Nanotechnology

6.1 Introduction

Nanomedicine and nano delivery systems are a relatively new but rapidly developing science where materials in the nanoscale range are employed to serve as means of diagnostic tools or to deliver therapeutic agents to specific targeted sites in a controlled manner. Nanotechnology offers multiple benefits in treating chronic human diseases by site-specific and target-oriented delivery of precise medicines. This chapter deals with recent advances in the field of nanomedicines and nano-based drug delivery systems/nanocarriers (NCs) through comprehensive scrutiny of the discovery and application of nanomaterials (NMs) in improving both the efficacy of novel and old drugs (e.g., natural products), selective diagnosis through disease marker molecules, and their toxicological potentials.

6.2 Delivery Routes

Drugs have long been used to improve health and extend lives. The practice of drug delivery has changed dramatically in the past few decades and even greater changes are anticipated in the near future. Targeted drug delivery, sometimes called smart drug delivery, is a method of delivering medication to a patient in a manner that increases the concentration of the medication in some parts of the body relative to others. Drugs can be introduced into the body via several different routes. These routes are generally classified by their "starting point"—the location at which the drug is administered. Each route has its own advantages and disadvantages. Different routes of drug delivery include:

- Buccal drug delivery.
- Nasal drug delivery.
- Ocular drug delivery.
- Oral drug delivery.
- Parenteral drug delivery (intramuscular, subcutaneous, intravenous, etc.)
- Pulmonary drug delivery.
- Sublingual drug delivery.
- Transdermal drug delivery.
- Vaginal/anal drug delivery.

Targeted drug delivery approaches are of great interest as they enable concentrated delivery of a drug compound to its desired target—increasing efficacy and reducing off-target effects. By modifying drug molecule properties, it is possible to optimize bioavailability, decrease clearance, and increase stability, making them ideal "carriers" for delivering a particular drug to its specific target-tissue. NPs have good solubility and consequently increased bioavailability due to their small size and larger surface area. Their appeal as drug carriers is enhanced by their ability to:

(a) cross the blood–brain barrier (BBB), (b) enter the pulmonary system, and (c) pass through the tight junctions of endothelial cells.

Nanoparticles have been explored as carriers for drugs to treat numerous conditions including cancer, neurological disorders, and acquired immune deficiency syndrome (AIDS). These NPs can enter the body via three main routes; injection, inhalation, or taken orally (see also Chap. 3). Nanotechnology-based formulations are largely parenteral, with some intended for oral administration. It is hoped that a significant number of preclinical and clinical trials would lead to the production of novel nanotherapeutics intended for non-parenteral delivery routes, such as pulmonary, nasal, vaginal, ocular, and dermal delivery routes. The efficacy of such particles can be achieved by changing the surface properties of the particle.

6.3 Nanocarriers

A NC is NP/NM being used as a transport module for another substance, such as a drug. NCs are currently being studied for their use in drug delivery and their unique characteristics that demonstrate potential use in chemotherapy. NMs consisting of polymeric nanostructures, organic, inorganic or metallic are frequently considered in designing the target-specific drug delivery systems. A tentative list of various types of NCs used for drug delivery in cancer therapy include: (a) lipid-based NCs, (b) inorganic NPs, and (c) polymeric NPs (Fig. 6.1). Those drugs having poor

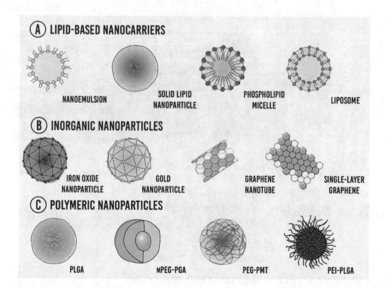

Fig. 6.1 Types of nanocarriers used for drug delivery in cancer therapy: (**a**) lipid-based nanocarriers, (**b**) inorganic NPs, and (**c**) polymeric NPs
https://www.mdpi.com/pharmaceutics/pharmaceutics-13-01167/article_deploy/html/images/pharmaceutics-13-01167-g001-550.jpg

solubility with less absorption ability are tagged with NPs/NMs. However, the efficacy of these nanostructures as drug delivery vehicles varies depending on the size,
shape, and other inherent biophysical/chemical characteristics. Lately, many
researchers have developed various NCs for liver tumor–targeted drug delivery and
imaging.

6.3.1 Organic NCs

Polymeric NMs exhibit characteristics ideal for an efficient delivery vehicle.
Because of their high biocompatibility and biodegradability properties, various
natural and synthetic polymers such as polyvinyl alcohol, poly-L-lactic acid, poly-
ethylene glycol, and poly (lactic-co-glycolic acid) and alginate and chitosan are
extensively used in the nanofabrication of NPs. Recently there have been enormous
developments in the field of drug delivery systems to provide active compounds to
its target location for treatment of various ailments. A number of drug delivery systems have been successfully employed; however, there are still certain challenges
that need to be addressed. An advanced technology needs to be developed for successful delivery of drugs to its target sites. Classification of biopolymers based upon
their origin that are used in nanomedicine obtained from different sources are summarized in Fig. 6.2. A few other natural bioactive materials used as NCs are discussed later in this chapter under Sect. 6.3.4.

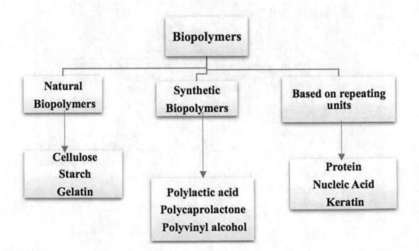

Fig. 6.2 Classification of biopolymers based upon their origin
https://www.mdpi.com/polymers/polymers-14-00983/article_deploy/html/images/
polymers-14-00983-g001-550.jpg

The success of these biopolymers in nanomedicine and drug delivery is due to their versatility and specified properties because they can originate from soft gels, flexible fibers, and hard shapes, so they can be porous or non-porous; they have great similarity with components of the extracellular matrix, which may be able to avoid immunological reactions. Biopolymeric NPs are biodegradable and gaining increased attention for their ability to serve as a viable carrier for site-specific delivery of vaccines, genes, drugs, and other biomolecules in the body. They offer enhanced biocompatibility, superior drug/vaccine encapsulation, and convenient release profiles for a number of drugs, vaccines, and biomolecules to be used in a variety of applications in the field of medicine. A few of the most extensively used biodegradable polymer matrices for preparation of NPs are:

(a) PLGA (poly-D-L-lactide-co-glycolide).
(b) PLA (poly-lactic acid).
(c) PCL (poly-ε-caprolactone).

These NPs have attracted considerable interest over recent years due to their properties resulting from their small size. Advantages of microspheres as drug carriers include: their potential use for controlled release, the ability to protect drugs and other molecules with biological activity against the environment, and the ability to improve their bioavailability and therapeutic index. They can be used for both localized and targeted delivery of drugs. The drug is dispersed throughout the polymer matrix of the microsphere. Microspheres can be prepared using either natural polymers or synthetic polymers. Synthetic polymers are either biodegradable or non-biodegradable. The persistence of non-biodegradable microspheres in the body can increase the risk of toxicity over longer time periods. Biodegradable polymers, however, do not pose the same risk, making them better suited to parenteral applications.

Advantages of polymeric NPs as drug carriers include their potential use for controlled release, the ability to protect drug and other molecules with biological activity against the environment, and the ability to improve their bioavailability and therapeutic index. Nanospheres are based on a continuous polymeric network in which the drug can be retained inside or adsorbed onto their surface. Schematic representation of the structure of nanocapsules and nanospheres used in drug delivery is shown in Fig. 6.3.

Depending upon the type of polymeric NPs/method used and the type of drugs/ bioactive ingredients loaded in polymeric NPs, these carriers can have different applications such as drug delivery, theranostics, or bioimaging, anti-glioma activity, anti-inflammatory activity, diabetic retinopathy, neovascular age-related macular degeneration (ocular neovascularization), anti-leishmanial (Leishmania infections), anti-fungal activity, etc.

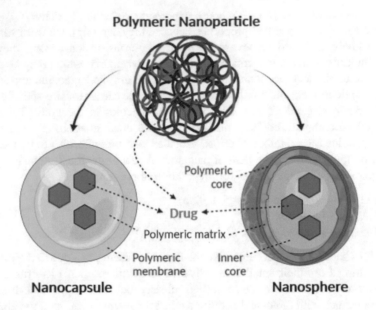

Fig. 6.3 Schematic representation of the structure of nanocapsules and nanospheres (*arrow* stands for the presence of drug/bioactive within the NPs)
https://www.mdpi.com/molecules/molecules-25-03731/article_deploy/html/images/molecules-25-03731-g001-550.jpg

6.3.1.1 Liposomes and Micelles

Liposomes are composed of a lipid bilayer separating an aqueous internal compartment from the bulk aqueous phase. Micelles are closed lipid monolayers with a fatty acid core and polar surface, or polar core with fatty acids on the surface (inverted micelle).

Liposomes Liposomes are spherical vesicles consisting of one (or more) phospholipid bilayers. These vesicles present themselves as an attractive delivery system due to them possessing flexible biochemical and physiochemical properties, allowing them to be easily manipulated. Liposomes have a unique ability to encase lipophilic and hydrophilic compounds, making them suitable carriers for a range of drugs. These NPs can be used in drug delivery for the treatment of tuberculosis (TB). The traditional treatment for TB is chemotherapy. However, chemotherapy over time has led to drug-resistant disease. The liposome delivery system of chemotherapy allows for better microphage penetration and better builds a concentration at the infection site. The delivery of the drugs works intravenously and by inhalation. Oral intake is not advised because the liposomes break down in the gastrointestinal system.

The advantages of liposomes include their capacity to self-assemble, ability to carry large drug-loads, and their biocompatibility. Liposomes can be easily manipulated and modified—with modification comes increased chances of detection and clearance. Being composed of natural phospholipids makes them "pharmacologically inactive," meaning they display minimal toxicity. Liposomes can be categorized into four main types such as conventional, theranostic, PEGylated, and ligand-targeted. Liposomes are generally used as NCs for (a) diagnostic applications to detect proteins, DNA, and small molecule targets using fluorescence, magnetic resonance, ultrasound, and nuclear imaging; (b) therapeutic applications based on small molecule-based therapy, gene therapy, and immunotherapy; and (c) theranostic applications for simultaneous detection and treatment of heavy metal toxicity and cancers.

Micelles Oral administration is the most commonly used and readily accepted form of drug delivery; however, it is observed that many drugs are difficult to attain enough bioavailability when administered via this route. Polymeric micelles (PMs) can overcome some limitations of the oral delivery acting as carriers because they are able to enhance drug absorption, by providing (1) protection of the loaded drug from the harsh environment of the GI tract, (2) release of the drug in a controlled manner at target sites, (3) prolongation of the residence time in the gut by mucoadhesion, and (4) inhibition of efflux pumps to improve the drug accumulation. Among these approaches, PMs have gained considerable attention in the last two decades as a multifunctional nanotechnology-based delivery system for poorly water-soluble drugs. Several physicochemical parameters seem to influence translocation of micelles across the epithelium, including surface hydrophobicity, polymer nature, and particle size.

Chitosan Chitosan exhibits mucoadhesive properties and can be used to act in the tight epithelial junctions. Thus, chitosan-based NMs are widely used for continued drug release systems for various types of epithelia, including buccal, intestinal, nasal, eye, and pulmonary. Similarly, NPs of carboxymethyl chitosan for the release of intra-nasal carbamazepine (CBZ) to bypass the BBB membrane have been used, thereby increasing the amount of the medication in the brain, refining the treatment efficacy, and thus reducing the systemic drug exposure.

Alginate Another biopolymeric material that has been used as a drug delivery is alginate. This biopolymer presents final carboxyl groups, being classified as anionic mucoadhesive polymer, and presents greater mucoadhesive strength when compared with cationic and neutral polymers.

Cellulose Cellulose and its derivatives are extensively utilized in the drug delivery systems basically for modification of the solubility and gelation of the drugs that resulted in the control of the release profile of the same. The presence of the hydrogen bonds between the cellulose nanocrystals and the drug results in sustained release of the same, and subsequently the NPs made with oxidized cellulose nanocrystals present lower release when compared to the NPs produced with native cellulose nanocrystals.

Gelatin Gelatin is extensively used in food and medical products and is a nontoxic alternative. Gelatin NPs are very efficient in delivery and controlled release of the drugs. They are nontoxic, biodegradable, bioactive, and inexpensive. Gelatin is a poly-ampholyte consisting of both cationic and anionic groups along with a hydrophilic group. It is known that the mechanical properties such as swelling behavior and thermal properties of gelatin NPs depend significantly on the degree of cross-linking between cationic and anionic groups. These properties of gelatin can be manipulated to prepare desired type of NPs from gelatin. Gelatin NPs can be prepared by the desolvation/coacervation or emulsion methods.

Protein and Polysaccharides Polysaccharides and proteins are collectively called as natural biopolymers and are extracted from biological sources such as plants, animals, microorganisms, and marine sources.

Protein-based NPs are generally decomposable, metabolizable, and are easy to functionalize for its attachment to specific drugs and other targeting ligands. They are normally produced by using two different systems: (a) from water-soluble proteins like bovine and human serum albumin and (b) from insoluble ones like zein and gliadin. The protein-based NPs are chemically altered in order to combine targeting ligands that identify exact cells and tissues to promote and augment their targeting mechanism.

The polysaccharides are composed of sugar units (monosaccharides) linked through O-glycosidic bonds. The composition of these monomers as well as their biological source are able to confer to these polysaccharides, a series of specific physical–chemical properties. One of the main drawbacks of the use of polysaccharides in the nanomedicine field is its degradation (oxidation) characteristics at high temperatures (above their melting point), which are often required in industrial processes. Besides, most of the polysaccharides are soluble in water, which limits their application in some fields of nanomedicine, such as tissue engineering.

Organic Nanofibers Organic nanofibers have been used as vectors for drug delivery for liver cancer. The chemotherapeutic drug-bearing nanofiber simultaneously inhibits metastasis and tumor growth of liver cancer cells.

6.3.1.2 Dendrimers

Dendrimers are highly bifurcated, monodisperse, well-defined, three-dimensional structures. They are globular-shaped and their surface is functionalized easily in a controlled way, which makes these structures excellent candidates as drug delivery agents. Dendrimers are grouped into several kinds according to their functionalization moieties: polyamidoamine (PAMAM), propylene imine (PPI), liquid crystalline, core–shell, chiral, peptide, glycodendrimers, and polyamidoamine-organosilicon (PAMAMOS). PAMAM is the most studied for oral drug delivery because it is water soluble and it can pass through the epithelial tissue boosting their transfer via the paracellular pathway. Dendrimers are limited in their clinical applications because of the presence of amine groups. These groups are positively charged or cationic which makes them toxic, hence dendrimers are usually modified in order to reduce this toxicity issue or to eliminate it.

6.3.1.3 Nanocrystals

Due to their particle size, nanocrystals (nanosuspension) have the great advantage of being intravenously injectable, reaching 100% bioavailability. Nanocrystals in the range of 100–300 nm can be injected intravenously without any unwanted effect, such as the obstruction of small capillaries. Consequently, NPs circulate in the bloodstream and dissolve according to their dissolution properties, and then are able to reach the target tissue. One of the most powerful applications of the intravenous injection of drug nanocrystal suspensions is the delivery of anticancer drugs.

6.3.2 Inorganic NCs

Inorganic NCs generally have physical properties, such as optical absorption, fluorescence (semiconductor QDs), and magnetic moment (e.g., iron oxides), useful reactive groups for different biomolecules in order to achieve a biological functionality, such as active targeting of tissues or cells. Some inorganic NCs, for example, iron, quantum dots (QDs), gold NPs, carbon-based NCs, and silica NPs, exhibit distinctive advantages for drug delivery, including high surface-to-volume ratio, controllable size and shape, and potential imaging functions. These NPs show several advantages such as good biocompatibility and versatility when it comes to surface functionalization. Biocompatibility, ease of synthesis, and ease of surface functionalization are among the significant properties of nonporous and mesoporous silica NPs in various nanomedicine applications. Due to their high brightness; long-lasting, wide, and continuous absorption spectra; and high fluorescent quantum yield, quantum dots are being used as the new optical probes for bioassays. Similarly, gold NPs provide ease of preparation, stability, low cytotoxicity, and high extinction coefficient of light from visible to "near infrared" regions, which are some properties that introduced them as important candidates in cancer drug and NC development. As a specific type of inorganic NMs, magnetic NPs that exhibit super paramagnetic are capable of being used as contrast agents in magnetic resonance imaging, site-specific gene and drug delivery, and diagnostic agents in the presence of an external magnetic field.

6.3.2.1   Graphene Oxide–Based NCs

Graphene oxide NCs have excellent drug loading capacity. The property of high electron transfer of individual graphene sheets makes them a good carrier for drug delivery application. Due to their high surface area, graphene-based NCs can form specific interactions with various drug molecules.

6.3.2.2   Nanoshells

Nanoshells have attracted tremendous attention over the past few decades as a promising tool for liver cancer therapy. It is a self-assembled polymer forming a core or shell structure and has been used for liver imaging as well. One of the most useful nanoshells is gold nanoshell. Gold nanoshell shows good targeting ability to liver cancer cells, for example, BEL7404 and BEL-7402, without affecting the normal healthy liver cell like HL-7702.

6.3.2.3   Carbon Nanotubes

Carbon nanotubes (CNTs) are needle-shaped materials that carry therapeutic drugs to the cellular component. CNTs are potentially considered excellent nano-vehicles for the.delivery of different therapeutic agents due to their small size and mass, high electrical, strong mechanical potency and thermal conductivity. CNTs exhibit biocompatibility, low toxicity, fewer side effects, and high treatment efficacy with low drug doses in tumor-targeted drug delivery.

6.3.3   Metallic NCs

The interest in using metallic NPs has been growing in different medical applications, such as bioimaging, biosensors, target/sustained drug delivery, hyperthermia, and photoablation therapy. In addition, the modification and functionalization of these NPs with specific functional groups allow them to bind to antibodies, drugs, and other ligands. Although the most extensively studied metallic NPs are gold, silver, iron, and copper, recently other kinds of metallic NPs such as zinc oxide, titanium oxide, platinum, selenium, gadolinium, palladium, cerium dioxide, and superparamagnetic iron-oxide NPs (SPIONs) have shown significant performance in the field of liver cancer therapy and diagnosis.

6.3.4 Natural NCs

Currently, the scientific community is focusing on the studies related to bioactive compounds, its chemical composition and pharmacological potential of various plant species, to produce innovative active ingredients that present relatively minor side effects than existing molecules. These plant- or microorganism-derived compounds have shown potential as therapeutic agents against various diseases such as cancer, microbial infection, and inflammation. However, their success in clinical trials has been less impressive, partly due to the compounds' low bioavailability. The incorporation of NPs into a delivery system for natural products would be a major advance in the efforts to increase their therapeutic effects. Recently, advances have been made showing that NPs can significantly increase the bioavailability of natural products both in vitro and in vivo. Nanotechnology has demonstrated its capability to manipulate particles in order to target specific areas of the body and control the release of drugs. Although there are many benefits to applying nanotechnology for better delivery of natural products, it is not without issues. In this sense, the accelerated development of nanotechnology has driven the design of new formulations through different approaches, such as driving the drug to the site of action (nanopharmaceutics), image and diagnosis (nanodiagnostic), medical implants (nanobiomaterials), and the combination of diagnosis and treatment of diseases (nanotheranostics). Thus, it is expected that the scientific development of nanotechnology can revolutionize the development of formulations based on natural products, bringing tools capable of solving various problems that limit the application of these compounds in large scale in nanomedicine. However, drug targeting remains a challenge and potential NP toxicity needs to be further investigated, especially if these systems are to be used to treat chronic human diseases such as cancer. Some examples of biological compounds obtained from higher plants and their uses in the field of nanomedicine are described in Fig. 6.4.

6.3.5 Hybrid NPs

Multiple nanocomponents with diagnostic and therapeutic functions can be integrated into a single nanosystem. These systems follow on the concept of a "theranostic" device, in which both diagnostic and therapeutic functions can be administered in a single dose. For NPs, one advantage of combining imaging with therapeutic functions is that the biodistribution of the materials can be monitored in vivo, reducing the potential for unintended side effects of drug toxicity or hyperthermia-induced damage in healthy tissues. In addition to the utility of tracking the fate of nanotherapeutics in vivo immediately after administration, the use of such hybrid NPs potentially allows the medical team to monitor the progress and efficacy of a therapy throughout the course of treatment.

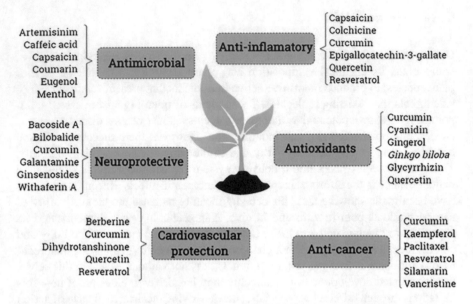

Fig. 6.4 Examples of natural compounds extracted from higher plants used in nanomedicine aiming at different approaches. Some of these extracts are already being marketed, others are in clinical trials, and others are being extensively studied by the scientific community
https://media.springernature.com/lw685/springer-static/image/art%3A10.1186%2Fs
12951-018-0392-8/MediaObjects/12951_2018_392_Fig5_HTML.png?as=webp

6.4   NCs for Theranostic Drug Delivery

Theranostic drug delivery literally means putting together diagnostic and therapeutic agents on a carrier to cure and diagnose cancer. Theranostic NCs have emerged to diagnose and treat the diseases at the cellular and molecular level. Figure 6.5 represents the theranostic model of NCs. Currently, the theranostic delivery–based approach has been explored effectively for treating liver cancer. The therapeutic agents in theranostic NCs include chemotherapeutics drugs, proteins, peptides, and gene and genetic materials. Diagnostic agents that are commonly used in theranostic NCs include gadolinium; fluorescent dyes; quantum dots; superparamagnetic iron oxides; radionuclides; and heavy elements such as iodine for optical imaging, magnetic resonance imaging (MRI), nuclear imaging, and computed tomography.

6.5   Factors Affecting Drug-Laden NCs

A drug's efficacy can be affected significantly by the way in which it is delivered. Using the best-suited delivery mechanism for a specific drug molecule it is possible to "optimize" the performance of that drug inside the body. Targeted drug delivery

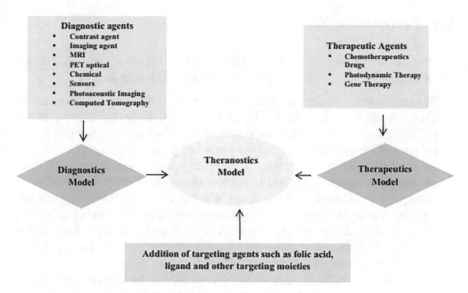

Fig. 6.5 Theranostics nanocarrier model
https://www.ncbi.nlm.nih.gov/pmc/articles/instance/7060777/bin/IJN-15-1437-g0002.jpg

approaches using NCs are of great interest as they enable concentrated delivery of a drug compound to its desired target—increasing efficacy and reducing off-target effects. For example, polymer-based NPs are submicron-sized polymeric colloidal particles in which a therapeutic agent of interest can be embedded or encapsulated within their polymeric matrix or adsorbed or conjugated onto the surface serve as an excellent vehicle for delivery of a number of biomolecules, drugs, genes, and vaccines to the site of interest in vivo. There is a size limit for the particles to be able to cross the intestinal mucosal barrier of the gastrointestinal (GI) tract after the drug has been delivered orally. Most often, macroparticles could not cross mucosal barrier due to their bigger sizes resulting in failed delivery of drugs. NPs on the other hand have an advantage over macroparticles due their nano-sizes. NPs, however, have a different set of problems of their own. They have a very short circulating life span within the body after intravenous administration. The NPs administered intravenously are rapidly cleared from the body by phagocytic cells; however, the problem of phagocytic removal of NPs has been solved by surface modification of NPs. Now, a wide variety of biomolecules, vaccines, and drugs can be delivered into the body using NCs and a number of routes of delivery. NPs can be used to safely and reliably deliver hydrophilic drugs, hydrophobic drugs, proteins, vaccines, and other biological macromolecules in the body. They can be specifically designed for targeted drug delivery to the brain, arterial walls, lungs, tumor cells, liver, and spleen. They can also be designed for long-term systemic circulation within the body. In addition, NPs tagged with imaging agents offer additional opportunities to exploit optical imaging or MRI in cancer diagnosis and guided hyperthermia therapy.

6.6   Properties of NCs

NP drug delivery focuses on maximizing drug efficacy and minimizing cytotoxicity. NP properties for effective drug delivery involves the following factors:

- The surface-area-to-volume ratio of NPs can be altered to allow for more ligand binding to the surface. Increasing ligand binding efficiency can decrease dosage and minimize NP toxicity. Minimizing dosage or dosage frequency also lowers the mass of NP per mass of drug, thus achieving greater efficiency.
- Surface functionalization of NPs is another important design aspect and is often accomplished by bioconjugation or passive adsorption of molecules onto the NP surface. By functionalizing NP surfaces with ligands that enhance drug binding, suppress immune response, or provide targeting/controlled release capabilities, both a greater efficacy and lower toxicity are achieved. Efficacy is increased as more drug is delivered to the target site, and toxic side effects are lowered by minimizing the total level of drug in the body.
- The composition of the NP can be chosen according to the target environment or desired effect. For example, liposome-based NPs can be biologically degraded after delivery, thus minimizing the risk of accumulation and toxicity after the therapeutic cargo has been released. Metal NPs, such as gold NPs, have optical qualities that allow for less invasive imaging techniques.
- The photothermal response of NPs to optical stimulation can be directly utilized for tumor therapy.

6.7   Uses and Applications

Nanocarrier-based targeted drug delivery can be used to treat many diseases, such as cardiovascular diseases and diabetes. However, the most important application of targeted drug delivery is to treat cancerous tumors. In doing so, the passive method of targeting tumors takes advantage of the enhanced permeability and retention (EPR) effect. This is a situation specific to tumors that results from rapidly forming blood vessels and poor lymphatic drainage. When the blood vessels form so rapidly, large fenestrae result that are 100–600 nm in size, which allows enhanced NP entry. Further, poor lymphatic drainage means that the large influx of NPs are rarely leaving, thus, the tumor retains more NPs for successful treatment to take place.

Due to increase in the number of cardiovascular diseases, diabetes, and cancerous tumors, there is a need to come up with an optimum recovery system. The key to solving this problem lies in the effective use of pharmaceutical drugs that can be targeted directly to the diseased tissue. This technique can help develop many more regenerative techniques to cure various diseases. The development of a number of regenerative strategies in recent years for curing heart disease represents a paradigm shift away from conventional approaches that aim to manage heart disease. Recent developments have shown that there are different endothelial surfaces in tumors, which has led to the concept of endothelial cell adhesion molecule-mediated targeted drug delivery to tumors.

6.8   Targeted Drug Delivery

Six methods in which NCs can deliver drugs include:

(a) Passive targeting.
(b) Active targeting.
(c) Vascular targeting (endothelial cells).
(d) Targeting the mildly acidic tumor microenvironment (ph specificity).
(e) Temperature specificity.
(f) Nuclear targeting.

6.8.1   Passive Targeting

Passive targeting refers to a NC's ability to travel down a tumor's vascular system, become trapped, and accumulate in the tumor. This accumulation is caused by the EPR effect which refers to the poly (ethylene oxide) (PEO) coating on the outside of many NCs. PEO allows NCs to travel through the leaky vasculature of a tumor, where they are unable to escape. The leaky vasculature of a tumor is the network of blood vessels that form in a tumor, which contain many small pores. These pores allow NCs in, but also contain many bends that allow the NCs to become trapped. As more NCs become trapped, the drug accumulates at the tumor site. This accumulation causes large doses of the drug to be delivered directly to the tumor site. PEO may also have some adverse effects on cell-NC interactions, weakening the effects of the drug, since many NCs must be incorporated into the cells before the drugs can be released. Since the encapsulation of small-molecule drugs in nanosized drug carriers enhances their pharmacokinetics (prolonged systemic circulation), this provides some tumor selectivity and decreases side effects. This type of tumor targeting termed "passive" relies on carrier characteristics (size, circulation time) and tumor biology (vascularity, leakiness, etc.), but does not possess a ligand for specific tissue or organ binding. Therefore, passive targeting cannot deliver large solutes and there arises the need for alternative tactics which has led to the development of other methods, such as active targeting.

6.8.2   Active Targeting

Active targeting involves the incorporation of targeting modules such as ligands or antibodies on the surface of NCs that are specific to certain types of cells around the body. NCs have such a high surface-area-to-volume ratio allowing for multiple ligands to be incorporated on their surfaces. These targeting modules allow for the NCs to be incorporated directly inside of cells, but also have some drawbacks. Ligands may cause NCs to become slightly more toxic due to non-specific binding, and positive charges on ligands may decrease drug delivery efficiency once inside of cells. However, active targeting has been shown to help overcome multi-drug resistance in tumor cells. Active targeting is able to significantly increase the quantity of drug delivered to the target cell compared to free drug or passively targeted nanosystems. After accumulation in the tumor region, the drug efficiency can be even increased. This is achieved through the decoration of the NC surfaces with ligands binding to receptors over-expressed onto the tumor cells. This strategy will improve the affinities of the NCs for the surface of cancer cell and thus enhance the drug penetration.

In active targeting, the drug-loaded NCs with receptor-specific ligands circulate in the blood stream and release the drug at the target tumor site. While in passive targeting, the EPR effect enables the drug conjugated NCs to penetrate deep into the tumor site (Fig. 6.6).

6.8.3   Vascular Targeting (Endothelial Cells)

Tumor blood vessels are mainly composed of endothelial cells and pericytes. A variety of angiogenic receptors are expressed on tumor-associated endothelial cells. Ligands, such as small molecules, peptides, aptamers, and antibodies, are chemically conjugated to the surface of NPs for the recognition of angiogenic markers. Antivascular agents, including chemotherapeutics, proteins, peptides, and nucleotides, can be encapsulated in the NC to achieve enhanced pharmacokinetics and efficacies.

In vascular targeting, the goal is to target angiogenic endothelial cells, which are adjacent to tumor cells and have intimate contact with blood vessels. This will reduce blood supply to the tumor and deprive cancer cells from oxygen and nutrients with subsequent hypoxia and necrosis. The mechanism involved in nanotechnology-mediated tumor vascular targeting is shown in Fig. 6.7. An important advantage of vascular targeting lies in the fact that its efficiency is not correlated to the specific blood vessel permeability or cell uptake. Vascular targeting is able to limit poor delivery of drugs and the drug resistance and can be more adapted to the tumor heterogeneity or to different sorts of tumors.

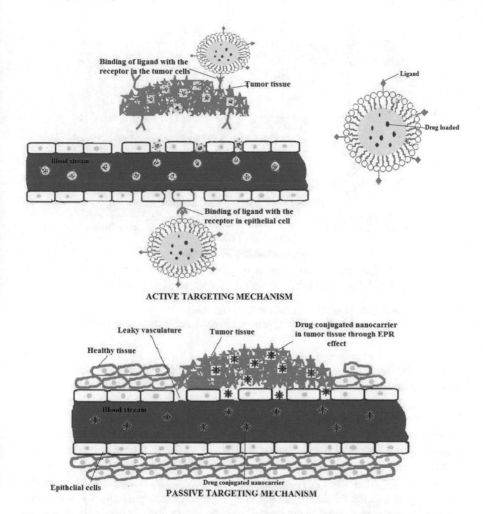

Fig. 6.6 Active vs. passive transport (mechanism of nanocarrier targeting). In active targeting, the drug-loaded NCs with receptor-specific ligands circulate in the blood stream and release the drug at the target tumor site. While in passive targeting, the EPR effect enables the drug conjugated NCs to penetrate deep into the tumor site
https://media.springernature.com/full/springer-static/image/art%3A10.1007%2Fs10311-018-00841-1/MediaObjects/10311_2018_841_Fig2_HTML.png?as = webp

6.8.4   Targeting the Mildly Acidic Tumor Microenvironment (pH Specificity)

Tumors are generally more acidic than normal human cells, with a pH around 6.8. Normal tissue has a pH of around 7.4. Certain NCs will only release the drugs they contain in specific pH ranges. pH specificity also allows NCs to deliver drugs directly to a tumor site. NCs that only release drugs at certain pH ranges can

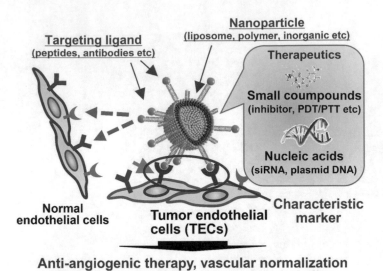

Fig. 6.7 Nanotechnology-mediated tumor vascular targeting
https://www.mdpi.com/ijms/ijms-20-05819/article_deploy/html/images/ijms-20-05819-ag.png

therefore be used to release the drug only within acidic tumor environments. High acidic environments cause the drug to be released due to the acidic environment, degrading the structure of the NC. These NCs will not release drugs in neutral or basic environments, effectively targeting the acidic environments of tumors while leaving normal body cells untouched. This pH sensitivity can also be induced in micelle systems by adding copolymer chains to micelles that have been determined to act in a pH-independent manner. These micelle-polymer complexes also help to prevent cancer cells from developing multi-drug resistance. The low pH environment triggers a quick release of the micelle polymers, causing a majority of the drug to be released at once, rather than gradually like other drug treatments. This quick release mechanism significantly decreases the time it takes for anticancer drugs to kill a tumor, effectively preventing the tumor from having time to undergo mutations that would render it drug resistant.

6.8.5 Temperature Specificity

Some NCs have also been shown to deliver drugs more effectively at certain temperatures. Since tumor temperatures are generally higher than temperatures throughout the rest of the body, around 40 °C, this temperature gradient helps act as safeguard for tumor-specific site delivery.

6.8.6 Nuclear Targeting

Beside the drug delivery to the tumor microenvironment or more precisely to the tumor cells, some treatments need an even more precise level which is the drug delivery at organelle level, for example, nucleus, lysosomes, mitochondria, or endoplasmic reticulum. In that way, the therapeutic response will be maximized, and their toxic side effects minimized. In the case of delivery of therapeutic genes, the target is the cell nucleus to exert their effects in correcting dysfunctional or missing genes. On the other hand, cancer cell nucleus can be targeted for a destroying effect. Indeed, the mechanism of action of most anticancer drugs, for example doxorubicin, involves oxidative DNA damage and topoisomerase II inhibition within the nucleus.

6.9 Nanotoxicity and Safety

Nanotechnology-based carriers is a rapidly growing field having potential applications in many areas. To study toxicity of NPs, different types of cell cultures, including cancer cell lines, have been employed as in vitro toxicity models. It has been generally agreed that NPs interfere with either assay materials or with detection systems. So far, toxicity data generated by employing such models are conflicting and inconsistent. The biggest challenge faced by the scientific community involved in drug development is to deliver safe and effective dosage of drugs without causing systemic toxicity. However, toxicity of NC systems involves physiological, physicochemical, and molecular considerations. NP exposures through the skin, the respiratory tract, the gastrointestinal tract, and the lymphatics have been described. Knowledge about their potential toxicity and health impact is essential before these NMs can be used in real clinical settings. Furthermore, the underlying interactions of these NMs with physiological fluids is a key feature of understanding their biological impact, and these interactions can perhaps be exploited to mitigate unwanted toxic effects. Some NC systems may induce cytotoxicity and/or genotoxicity, whereas their antigenicity is still not well understood (see also Chap. 3). NCs may alter the physicochemical properties of xenobiotics resulting in pharmaceutical changes in stability, solubility, and pharmacokinetic disposition. In particular, NCs may reduce the toxicity of hydrophobic cancer drugs that are solubilized. Therefore, on the basis of available experimental models, it may be difficult to judge and list some of the more valuable NPs as more toxic to biological systems and vice versa. Considering the potential applications of NPs in many fields and the growing apprehensions of the FDA about the toxic potential of nanoproducts, it is the need of the hour to look for new internationally agreed free of bias toxicological models by focusing more on in vivo studies.

6.10 Challenges and Opportunities

The therapeutic potential and clinical application of macromolecules is hampered by various obstacles including their large size, short in vivo half-life, phagocytic clearance, poor membrane permeability, and structural instability. These challenges have encouraged researchers to develop novel strategies for effective delivery of macromolecules. Due to powerful breakthroughs in nanotechnology, smart delivery mechanisms have rapidly emerged for use in diverse applications across biomedical research and therapeutic development. Because disease targets for therapy are often localized in specific cells, organs, or tissues, an EPR-based strategy remains inadequate for accurate drug delivery and release to target regions, resulting in an insufficient drug concentration reaching the target region and undesired side effects. To address these issues, more precise and remote-controlled stimuli-responsive systems, which recognize and react to changes in the pathophysiological microenvironment, are being elucidated as feasible use of NCs for the drug delivery. In this perspective, chemical, physical, and biological engineering principles and strategies for constructing DNA-assisted NCs or other suitable techniques can be precisely reconfigured by external and internal stimuli to drive the release of a loaded drug in a target region with appropriate microenvironments. To do so, scientists need to guide the development of future in vivo therapies and clinical translation strategies.

Further Reading

Aleksandra Z, Filipa C, Oliveira AM, Neves A, et al. Polymeric nanoparticles: production, characterization. Toxicol Ecotoxicol Mol. 2020;25(16):3731. https://doi.org/10.3390/molecules25163731.

Chenthamara D, Subramaniam S, Ramakrishnan SG, et al. Therapeutic efficacy of nanoparticles and routes of administration. Biomater Res. 2019;23:20. https://doi.org/10.1186/s40824-019-0166-x.

Gupta PK. Chapter 14: Toxicology of nanoparticles. In: Problem solving questions in toxicology – a study guide for the board and other examinations. 1st ed. Cham: Springer; 2020a.

Gupta PK. Toxic effects of nanoparticles. In: Brain storming questions in toxicology. 1st ed. Boca Raton: Taylor & Francis Group, LLC, CRC Press; 2020b. p. 297–300.

Gupta PK. Chapter 15: Toxic effects of nanoparticles. In: Toxicology: resource for self study questions. 3rd ed. Seattle: Kinder Direct Publications; 2022a.

Gupta PK. Fundamentals of nanotoxicology. 1st ed. New York: Elsevier Inc.; 2022.

Mohamed FA, Nicolas A, Justine W, Ziad O, Thierry FV. An overview of active and passive targeting strategies to improve the NCs efficiency to tumor sites. J Pharm Pharmacol. 2019;71(8):1185–98. https://doi.org/10.1111/jphp.13098.

Patra JK, Das G, Fraceto LF, et al. Nano based drug delivery systems: recent developments and future prospects. J Nanobiotechnol. 2018;16:article 71. https://doi.org/10.1186/s12951-018-0392-8.

Hamid R, Manzoor I. Nanomedicines: nano based drug delivery systems challenges and opportunities [Online first]. IntechOpen; 2020. https://doi.org/10.5772/intechopen.94353. https://www.intechopen.com/online-first/nanomedicines-nano-based-drug-delivery-systems-challenges-and-opportunities

Ruman U, Fakurazi S, Masarudin MJ, Hussein MZ. NC-based therapeutics and theranostics drug delivery systems for next generation of liver cancer nanodrug modalities. Int J Nanomedicine. 2020;15:1437–56. https://doi.org/10.2147/IJN.S236927. https://biomaterialsres.biomedcentral.com/articles/10.1186/s40824-019-0166-x#Sec20

Chapter 7
Applications of Nanotechnology in Dentistry

Abstract Dental oral tissue is facing significant changes in clinical treatments in dentistry. Different studies are present in the literature about the use of nanomaterials in dentistry, and most of these studies are animal trials and only a few of them are about human patients. This chapter starts with a micro and macro morphological description of dental tissues, with a focus on the main diseases that may affect them, and discusses the classification of conventional and unconventional dental nanomaterials. Their applications that can be clinically resolved with the use of nanotechnology in prosthodontics, oral and maxillofacial surgery, conservative dentistry and endodontics, restorative dentistry, orthodontics and dentofacial orthopedics, oral medicine and radiology, preventive dentistry, dental implants, and dentin hypersensitivity have been summarized. Finally, the benefits and potentials of nanotoxicity and health implications have been discussed briefly.

Keywords Nanomaterials · Nano-Fillers · Dental Nanocomposite · Nano-Biotechnology · Nanodiagnostics · Dentistry · Dental diseases · Nanocarriers · Nanocomposites · Nanorobots · Prosthodontics · Nanotoxicity · Applications

7.1 Introduction

Nanomaterials (NMs) were used in dentistry for the first time in 2002 with the inclusion of nanofillers in composite resins for dental reconstruction. These NMs have been used in the prevention of main oral and dental biofilm-dependent diseases, like caries and periodontal diseases, with the addition of antibacterial and anti-demineralizing particles in toothpaste, mouthwashes, composite resins, and dental adhesive. This chapter deals with the classification of conventional and unconventional dental NMs. Their applications that can be clinically resolved with the use of nanotechnology in prosthodontics, oral and maxillofacial surgery, conservative dentistry and endodontics, restorative dentistry, orthodontics and dentofacial orthopedics, oral medicine and radiology, preventive dentistry, dental implants, dentin hypersensitivity, and their toxicological implications have been summarized.

© The Author(s), under exclusive license to Springer Nature Switzerland AG 2023
P K Gupta, *Nanotoxicology in Nanobiomedicine*,
https://doi.org/10.1007/978-3-031-24287-8_7

7.2   Structure and Periodontal Tissues

From a histological point of view, teeth are made of three strong tissues:

(a) Enamel
(b) Dentin
(c) Cementum

The dental pulp, in the middle, is responsible for the tropism of the dentin (Fig 7.1). The three hard tissues have a similar composition of inorganic components but with different percentages, because the enamel is fundamentally inorganic, while dentin and cement have an important organic component, and dental pulp is only organic. From an embryological point of view, enamel has an epithelial–ectodermal origin, but its forming cells (ameloblasts), once completed their function, undergo apoptosis and they are not present in adult tissue (except for some epithelial cells in periodontal tissues called epithelial rests of Malassez [ERM]). Instead, dentin and dental pulp (which together form the pulpodentinal complex) and the periodontal apparatus (made of cement, periodontal ligament, PDL), and alveolar bone are of mesenchymal origin.

Enamel is the hardest tissue of the human body, and it is acellular and translucent and forms the external surface of the teeth. It has different thicknesses, maximum in the cusps and minimum in the cementoenamel junction.

The main diseases of dental interest that may affect dental elements and their support apparatus are:

• Damage to the hard tissues of the tooth like caries, fractures, and cervical erosions, without loss of pulpal functionality

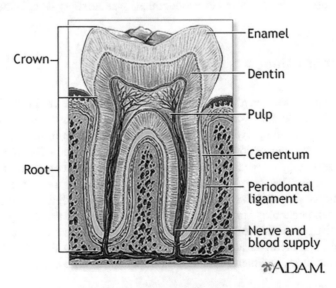

Fig. 7.1   Structure of teeth. (https://medlineplus.gov/ency/images/ency/fullsize/1121.jpg)

- Damage to the pulpodentinal complex with loss of pulpar vitality and, consequently, with removal by endodontic therapy
- Damage to periodontal complex and alveolar bone from periodontal disease and trauma
- Complete loss of one or more teeth in the most severe forms of these diseases or their absence for agenesis
- Small or medium bone losses by mandibular or maxillary cysts or odontogenic tumors

7.3 Classification of Dental NMs

Based on shape and composition of NPs/NMs used for dentistry include:

(a) Conventional NPs include metallic NPs and metal oxide NPs.
(b) Unconventional NPs include nano diamonds, nanoshells, and quantum dots. These are the new age fillers for advanced dental materials and they can be modified easily for the application.

7.3.1 Conventional NPs

7.3.1.1 Metallic NPs

Silver NPs (AgNPs) Antibacterial properties of Silver have been known for centuries. Additionally, it has also shown to have antiviral and anti-fungal properties. AgNPs are widely used in water purification, toothpaste, washing machines, shampoo, kitchen utilities, fabrics, nursing bottles, humidifiers, several areas of dentistry which includes endodontics, dental restorative material, dental prosthetics, dental implants. Incorporation of AgNPs decreases microbial colonization over dental parts and increases oral health because it can penetrate easily the bacterial cell membrane resulting in rapid bactericidal activity.

Gold NPs (AuNPs) Gold has inert and biocompatible behavior along with antibacterial properties. AuNPs have been introduced as a potential nano drug delivery carrier for cancer treatment and diagnosis. Recently, various Gold nano structures such as Gold nanospheres and nanorods has been used as photo thermal agents, contrast agents and nano drug delivery carriers.

Copper NPs (CuNPs) CuNPs provide sufficient antimicrobial activity against *S. Mutans* to the dental adhesives and prevent the degradation of the adhesive interface, without reduction in the mechanical properties of the formulations. Copper produces hydroxyl radicals that cause imbalance in the cell membrane of microorganisms which leads to leakage and cell death.

7.3.1.2   Metal Oxide NPs

Metal Oxide NPs are oxides of metal particles that are more stable in the oxide form. For example, Zinc Oxide NPs (ZnO NPs) have effective antibacterial activity at the nanoscale, and their antibacterial efficiency increases significantly. Titanium alloys are broadly used in dentistry as a result of their excellent properties such as high strength, high corrosion resistance, and good biocompatibility. Titanium Dioxide (TiO_2) has a catalytic activity and exposure to ultraviolet (UV) rays leads to generation of reactive oxygen species (ROS). This catalytic activity of TiO_2 causes an osmotic imbalance in microorganisms and interferes with their phosphorylation, leading to bacterial cell lysis. Zirconium dioxide (ZrO_2) has played a very important role in the manufacturing of dental materials. Zirconia, also known as "ceramic steel," has superior toughness, strength, fatigue resistance, excellent wear resistance, and biocompatibility. It has also shown very similar properties and colors to mimic the natural tooth, thereby ensuring better aesthetics. Alumina ceramics show better aesthetics, polished surface, wear resistance, and hardness, along with high biocompatibility to surrounding oral tissues. Dental materials that are lacking in mechanical strength can be made tougher through the addition of Al2O3 NP. Silicon dioxide (SiO2) NPs are used as a filler for dental restorative materials to improve the mechanical properties and fine powder of Silica as a polishing agent for polishing the rough surface of the tooth in order to prevent food accumulation or plaque formation.

7.3.2   Unconventional NMs

The prominent challenges in the development of advanced dental materials are to produce strong, nontoxic, and antibacterial materials by using various NPs. In this respect, the field of nanotechnology enables exploring various NPs to use and test for dental applications. Depending on the properties of the particles, various approaches for NP integration in dental materials have been implemented.

Nanodiamonds (NDs) Nanodiamonds are well-suited for biomedical use because they are carbon based and carbon is not toxic to the body. Recently, amoxicillin loaded ND-Gutta Percha composite (NDGP-AMC) has been developed for its use in root canal treatments with promising results. NDs have great potential for its use in various applications in restorative dentistry.

Quantum Dots Quantum dots are a semi-conductive NPs such as lead sulfide, zinc sulfide and indium sulfide that can emit light based on the amount and wavelength of light irradiated to them. Quantum dots are used for labeling live biological material. they can be injected into cells or attached to proteins in order to track, label or identify specific biomolecules. Quantum dots can be used for treatment of head and neck diseases via drug delivery and correction of genetic defects. They may also play a role in prevention of oral cancer.

Nanoshells Nanoshells are spherical particles consisting of a dielectric core surrounded by a thin metallic shell, most commonly gold. Because of their optical and chemical properties, these particles have been used for biomedical imaging and cancer treatment. NanoShells can be used for various therapeutic applications in dentistry.

Quaternary Ammonium Polyethyleneimine NPs (QPEI-NPs) QPEI-NPs is a crosslinked NP formulation containing quaternary ammonium polyethylenimine (QPPEI) with potential antibacterial activity. The cationic polymer PEI kills bacteria by rupturing their cell membranes without the development of resistance.

Quaternary Ammonium Methacrylate NPs (QAM-NPs) Many quaternary ammonium compounds have been synthesized that are not only antibacterial, but also possess antifungal, antiviral and anti-matrix metalloproteinase capabilities. Incorporation of quaternary ammonium moieties into polymers represents one of the most promising strategies for preparation of antimicrobial biomaterials. QAM resins have also shown effective bond strength properties with restorative dental adhesive materials, and their antibacterial ability may be useful for developing more advanced dental adhesive restorative materials.

Amorphous Calcium Phosphate NPs **(ACP-NPs)**Amorphous Calcium Phosphate NPs (ACP-NPs) in a chitosan aqueous solution in order to prevent and treat the existing enamel decalcification around the brackets and gingivitis. Incorporation of NPs in restorative dental materials can serve the dual action of having antibacterial effects and remineralization of the decayed tooth. In dental composites, the NPs of amorphous calcium phosphate releases calcium and phosphate ions around the restoration and help remineralization of the decayed tooth.

Carbon Nanotube (CNT) Carbon nanotubes (CNTs) have unique electrical as well as mechanical properties. Strength and flexibility of CNTs are because of C22C covalent bond and hexagonal orientation. CNTs also have thermal and electrical conductivity (semi conductivity). Because of it's excellent mechanical and electrical properties such as heat stability, heat transmission efficiency, high strength and lower density, it is used as a candidate for teeth filling and various applications.

Halloysite Nano-tube (HNT) Halloysite nanotubes (HNT) are a naturally occurring aluminosilicate with a predominantly hollow tubular structure mined from natural deposits. They have a natural milky white color with high strength and elastic modulus thereby making them one of the ideal fillers for dental composite fabrication. HNTs can also be loaded with antibiotic agents to develop a nano-filler with the ability to prevent the formation of oral bio-film and thereby preventing the development of secondary dental caries.

Nanoplatelets-Based NMs Nanoplatelets have naturally occurring functional groups like ethers, carboxyls, or hydroxyls located at the edges of the particles. Graphene is a prime candidate for nanoplatelet-based NMs. Graphene nanoplatelet NPs are comprised of short stacks of platelet-shaped graphene sheets that are identical to those found in the walls of carbon nanotubes, but in a planar form. Graphene oxide nanoplatelets are used in dentistry for their unique properties.

7.4   Dental Applications

There are a number of possible options to make smart materials for dental application. Some of the options for the production of smart materials for dental applications include:

(a) Material Synthesis: Producing synthetic materials matching morphology and properties similar to natural dental tissues.
(b) Biomimetic Approaches: To replace lost dental tissues follow nature's principles and produce biomaterials resembling their properties very close to the replacing tissues.
(c) Tissue Engineering: Use of regenerative medicine and tissue engineering approaches for replacing the lost dental tissues by regenerations.

Because of the growing interest in the future of the dental application of nanotechnology, a new field called nanodentistry is emerging. The development of nanodentistry will allow nearly perfect oral health by the use of NMs and biotechnologies including tissue engineering and nanorobots. With the development of new techniques, the use of NMs in dentistry has made a huge impact and opened new ways for various applications (Fig. 7.2). Disease conditions and application of NPs used in dentistry are summarized in Table 7.1.

7.4.1   Periodontics

Periodontists often treat more problematic periodontal cases, such as those with severe gum disease or a complex medical history. Periodontists offer a wide range of treatments, such as scaling and root planning (in which the infected surface of the root is cleaned) or root surface debridement (in which damaged tissue is removed). They can also treat patients with severe gum problems using a range of surgical procedures. In addition, periodontists are specially trained in the placement, maintenance, and repair of dental implants. Various treatments are available for the

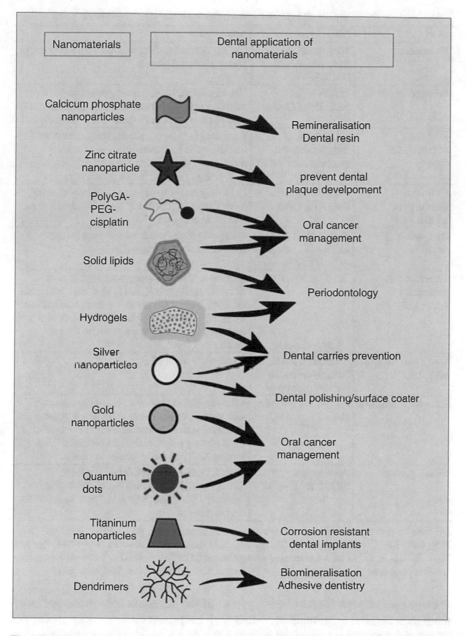

Fig 7.2 Various applications of Nanomaterials in Dentistry. (https://media.springernature.com/lw685/springer-static/image/chp%3A10.1007%2F978-981-13-8954-2_3/MediaObjects/474017_1_En_3_Fig4_HTML.jpg)

Table 7.1 Disease conditions and application of NPs used in different in dentistry

Discipline	NPs	Applications
Periodontics	Triclosan NPs, tetracycline NPs, HA NPs	Use in periodontal diseases and in the treatment of deep periodontal pockets.
Prosthodontics	ZrO2NPs, Al2O3NPs, CoCr NPs,	Nanoapatite for biofilm management on the tooth surface, remineralization. Nanocomposite surface coatings to prevent pathogenic bacteria adherence.
Oral and Maxillofacial surgery	HA NPs, Au NPs, carbon nanotube, quantum dots (QDs)	Used as optical probes for the early detection of oral cancer and as scaffolds for new bone formation. Quantum dots are used as alternative contrasting agents for the diagnosis of cancer.
Conservative dentistry and endodontics	Au NPs, Ag NPs, ZnO NPs, QPEI NPs	Used drug and gene delivery, tissue engineering, as an antibacterial agent in root canal sealers and for root canal disinfection.
Restorative dentistry	Ag NP, ZnO NPs, QPEI NPs, HA NPs, ACPNPs NPs	Composite resins and nanofillers are used as remineralizing, bonding agents.
Orthodontics and dentofacial orthopedics	Alumina NPs ZnO NPs	Treatment of dental hypersensitivity, coating of metal NPs on brackets, and arch wires to decrease the surface irregularities and reduction in friction. Metal NP coating acts as a solid lubricant film to ease the sliding of orthodontics wire over the bracket
Oral medicine and radiology	Silica NPs, ZrO2 NPs, HA NPs, TiO2 NPs quantum dots	Therapeutic application.
Preventive dentistry	ACPNPs, ZnO NPs, HA NPs	Used for final rinsing of root canal treatment. Low surface tension allows NPs to get to the smallest fissures and dental ducts of the system. Used directly before embedding dental fillings, and surface modifications of dental implants.
Dental implants	Au NPs, ZnO NPs, TiO2 NPs, CaP NPs	Surface modifications of dental implants, CaP NP are deposited on a double acid-etched surface by Discrete Crystalline Deposition (DCD) sol-gel process.

HA NPs Hydroxyapatite NPs, *ZrO₂ NPs* Zirconium dioxide NPs, *Al₂O₃ NPs* Aluminum Oxide NPs, *CoCr NPs* Cobalt Chromium NPs, *Au NPs* Gold nanoparticles, *Ag NPs* Silver NPs, *Zno NPs* Zinc Oxide NPs, *QPEI NPs* Quaternary ammonium polyethylenimine NPs, *ACPNPs* Amorphous calcium phosphate NPs, *TiO₂ NPs* Titanium dioxide, *CaP NPs* Calcium phosphate NPs

management of periodontal diseases; some of them include both medicinal treatment and surgical interventions. Medicinal treatments include drug molecules which are naturally macro-sized particles that find it challenging to penetrate the periodontal pockets. Conversely, the nanoscale sizes of NPs make it easier for them to reach subgingival regions.

7.4.2 Prosthodontics

Prosthodontics is a recognized dental specialty pertaining to the diagnosis, treatment planning, rehabilitation, and maintenance of the oral function, comfort, appearance, and health of patients with clinical conditions associated with missing or deficient teeth and/or oral and maxillofacial tissues. Many different metal combinations have been used for making a prosthodontic denture with titanium, cobalt-chromium, and molybdenum alloys. Their combinations show superior mechanical properties and corrosion resistance to stainless steel or gold alloy. Metal NPs such as HA NPs and TiO_2 NPs have better biological acceptance than traditional metals in the fabrication of prosthodontic dentures.

7.4.3 Oral and Maxillofacial Surgery

Maxillofacial surgery is a specialty that combines surgical training with dental expertise to correct a wide spectrum of diseases, injuries, tumors, defects, and deformities in the mouth, head, neck, face, jaws, and the hard and soft tissues of the oral and maxillofacial region. Most commonly used NMs include Alumina NPs and ZnO NPs. These NMs demonstrate excellent biocompatibility and provide superior results over traditional treatment options. They can also be used as scaffolds for new bone formation due to their ability to promote the osteogenic differentiation and biomineralization of cells. Localized nanodrug delivery can also help in preserving surrounding healthy tissues while targeting malignant tissue. NPs having magnetic NPs could be used for tumor-targeted drug delivery therapy. These NPs /NMs can be directly injected through an intravenous route to target the tumor site tissue, and due to their nano size, a low dosage of drugs that reduces the systemic toxicity is required, giving the desired effect of tumor regression through precise targeting of drug delivery.

7.4.4 Conservative Dentistry and Endodontics

Conservative dentistry is the branch of dentistry which is concerned with the conservation of teeth in the mouth. It embraces the practice of operative dentistry and endodontics, and includes various kinds of direct and indirect restorations of individual teeth in the mouth. Microorganisms in the oral cavity can cause dental caries that lead to various endodontic procedures such as root canal treatment, one of the important reasons for deep dental caries. Commonly used NMs include Au NPs, Ag NPs, ZnO NPs, and QPEI NPs. The inclusion of biopolymeric NPs in root canal disinfectants provided significant antibacterial activity, and the inclusion of QPEI

NPs improves the antibacterial activity of the root canal sealer against biofilms of *E. fecalis* (*Enterococcus fecalis*) strains.

7.4.5   Restorative Dentistry

Restorative dentistry refers to any dental procedure that repairs or replaces a tooth. Restorative procedures include cavity fillings, root canals, and even dental implants. There can be two goals in restorative dentistry: to restore the function of the teeth and to restore the appearance of the teeth. Nanotechnology has helped significantly in manufacturing of biocompatible and non-toxic dental restorative materials such as Ag NP, ZnO NPs, QPEI NPs, HA NPs, ACPNPs, GIC, dental composite, dental implants, and endodontic materials. The resin-based dental restorative materials have shown great advances in the last few years. The addition of NPs in the dental composite resin matrix can significantly increase its mechanical properties, low polymerization shrinkage with high abrasion resistance and surface hardness. Recently, nanoionomers (Nano; 3 M ESPE) have been marketed for clinical use.

7.4.6   Orthodontics and Dentofacial Orthopedics

Orthodontics and dentofacial orthopedics is the field of dentistry that primarily deals with the diagnosis, prevention, and correction of malpositioned teeth. Orthodontists help patients who have misaligned teeth by applying braces to straighten the gapped, crooked, and crowded teeth. With this treatment, patients can have a better bite and smiles. Friction and mechanical resistance between orthodontic wires and brackets can be reduced by coating with NPs. Nano coating of antibacterial NPs in orthodontic materials can prevent dental plaque formation around the orthodontic appliances and prevent dental caries associated with orthodontic treatments. Alumina, ZnO NPs, and inorganic fullerene-like NPs of tungsten sulfide (IF-WS2), which are potent dry lubricants, have been used as self-lubricating coatings for orthodontic stainless steel wires. These are capable of reducing friction between two surfaces sliding against each other without the need for a liquid media.

7.4.7   Oral Medicine and Radiology

Oral medicine and radiology is the specialty that focuses on the diagnosis and medical management of complex diagnostic and medical disorders affecting the mouth and jaws, and radiology is the field of diagnosis that uses conventional and advanced imaging methods. Organic and inorganic NPs of silica, zirconia, HA, and titanium dioxide have been used in oral medicine for therapeutic application. Digital imaging is tested with nano phosphor scintillators that rapidly emit visible light when exposed to even a very low dose of ionizing radiation. This nano-imaging technique requires a very low amount of radiation to give high-quality images compared to conventional methods, which can be very beneficial to dental applications.

7.4.8   Preventive Dentistry

Preventive dentistry is dental care that helps maintain good oral health. It's a combination of regular dental check-ups along with developing good habits like brushing and flossing. The inclusion of nanoapatite particles and dental restorative materials can facilitate the remineralization of damaged tooth structures, and antibacterial NPs in dental nanocomposites can prevent the pathogenic bacterial adhesion on the tooth surface. For example, dentifrices containing HA NPs can be used for bio-film reduction and remineralization of enamel lesions and incorporation of amorphous calcium phosphate NPs (ACPNPs) in the dental composite. This can also facilitate remineralization of tooth minerals.

7.4.9   Dental Implant

A *dental implant* is a surgical fixture that is placed into the jawbone and allowed to fuse with the bone over the span of a few months. Nanotechnology has provided opportunities for the manipulation of implant surfaces in its capacity to mimic the surface topography formed by extracellular matrix components of natural tissue. The possibilities introduced by nanotechnology now permit the tailoring of implant chemistry and structure with an unprecedented degree of control. The use of Au NPs, ZnO NPs, TiO_2 NPs, and CaP NPs provides new tools and capabilities that will result in faster bone formation, reduced healing time, and rapid recovery to function.

7.4.10   Dentin Hypersensitivity

Natural hypersensitive teeth have a much higher surface density of dentinal tubules and diameter in comparison to their nonsensitive counterparts. Reconstructive dental nanorobots, using native biological materials, could selectively and precisely occlude the dentinal tubules within minutes, offering patients a quick and permanent cure from dentin hypersensitivity.

7.5   Nanotoxicity of Dental NMs

In recent years, NMs have been widely used in the production of dental materials such as dental composites, root canal sealers, oral disease preventive drugs, prostheses, and for teeth implantation. NMs further carry oral fluid or drugs, maintain oral hygiene, prevent oral disease (cancer), and hence restore oral health care up to a high extent. However, the dental applications of NMs yield not only a significant improvement in clinical treatments but also growing concerns regarding their biosecurity. Because the central nervous system (CNS) may be a potential target organ of NMs, it is essential to determine the neurotoxic effects of NPs. Although the impact of NPs on the CNS has received considerable attention in recent years, the data and findings obtained from the in vivo and in vitro studies are still limited. Better testing and evaluation systems are urgently needed.

7.6   Conclusion

Recent developments and achievements in NMs and nanotechnology have provided a promising hand in the commercial applications of NMs in the diagnosis and management of periodontal diseases. Although many studies have been published concerning nanocomposite and nanoporous materials, it will become of increasing importance to specifically develop NMs for the management of periodontal diseases. It is envisaged that this trend will be further improved in the future as more and more nanotechnologies are commercially explored. At the same time, these dental NMs have created more exposure opportunities for both dental staff and patients. Dental applications of NMs yield not only a significant improvement in clinical treatments but also growing concerns regarding their biosecurity. Because the CNS may be a potential target organ of NMs, it is essential to determine the neurotoxic effects of NPs. The limitations of the present testing methods and the experimental models also make it difficult to establish a science-based evaluation system. In conclusion, more efforts are required to ensure the safe use of NMs.

Further Reading

Barot TD, Rawtani D and Kulkarni P. anotechnology-based materials as emerging trends for dental applications. Published by De Gruyter Open Access on April 17. 2021. https://doi.org/10.1515/rams-2020-0052
Elisa B, Mele S, Lia R. In: Mukherjee A, editor. Dental tissue engineering: A new approach to dental tissue reconstruction, biomimetics learning from nature. IntechOpen; 2010. https://doi.org/10.5772/8795.
Tatsuhiro F, Seiko T, Yusuke T, Reiko T-T, Satomura K. Dental pulp stem cell-derived, scaffold-free constructs for bone regeneration. Int J Mol Sci. 2018;19(7):1846. https://doi.org/10.3390/ijms19071846.

Gupta PK. Toxic effects of nanoparticles. In: Toxicology: resource for self study questions. 2nd ed. (Chapter 15),. Kinder Direct Publications; 2020a.

Gupta PK. Toxicology of nanomaterials. In: Problem solving questions in toxicology - a study guide for the board and other examinations. Ist ed, Chapter 14. Switzerland: Springer Nature; 2020b.

Gupta PK. Toxic effects of nanoparticles. In: Brain storming questions in toxicology. Ist ed. Taylor & Francis Group, LLC. CRC Press; 2020c. p. 297–300.

Gupta PK. Fundamentals of Nanotoxicology. 1st ed. USA: Elsevier Inc.; 2022.

Chieruzzi M, Pagano S, Moretti S, Pinna R, Milia E, Torre L, Stefano E. NMs for tissue engineering in dentistry. NMs (Basel). 2016;6(7):134. https://doi.org/10.3390/nano6070134. https://www.ncbi.nlm.nih.gov/pmc/articles/PMC5224610/

Khurshid Z, Zafar M, Qasim S, Shahab S, Naseem M, Reqaiba AA. Advances in nanotechnology for restorative dentistry. Materials (Basel). 2015;8(2):717–31. https://doi.org/10.3390/ma8020717. https://www.ncbi.nlm.nih.gov/pmc/articles/PMC5455275/

Chapter 8
Tissue Engineering and Regenerative Medicine

Abstract Tissue engineering and regenerative medicine (TERM) is a biomedical engineering discipline that uses a combination of cells, engineering, materials, methods, and suitable biochemical and physicochemical factors to restore, maintain, improve, or replace different types of biological tissues. TERM often involves the use of cells placed on tissue scaffolds in the formation of new viable tissue for a medical purpose but is not limited to applications involving cells and tissue scaffolds. This chapter covers the rationale of tissue engineering; the latest innovations; applications of nanoparticles; use of nanomaterials; delivery of bioactive agents/molecules (growth factors, chemokines, inhibitors, cytokines, genes, etc.); use of contrast agents in a controlled manner; their biocompatibility; toxicity; and safety aspects that are important implements to exert control over and monitor during TERM.

Keywords Hydrogel · Nanocomposite · Tissue engineering · Regenerative medicine · Bioactive agents · Delivery systems · Nanoparticles · Scaffolds · Growth factors · Nanomaterials · Biocompatibility · Nanotoxicity · Safety of nanoparticles · Toxicity · Applications

8.1 Introduction

There is an increasing demand for tissue engineering and regenerative medicine (TERM) solutions, which is a rapidly growing multidisciplinary field. The impact of nanotechnology has altered traditional and simple approaches in TERM toward more complex and efficient systems. TERM is an interdisciplinary field dedicated to the regeneration of functional human tissues. Despite the body having intrinsic self-healing properties, the extent of repair varies among different tissues, and may also be undermined by the severity of injury or disease. Therefore, the management of the health of tissues is a prime requirement in TERM. Along with nanoparticles

(NPs), other products of nanoscale technology such as nanofibers and nano-patterned surfaces have been used for directing cell behavior in the TERM field. This chapter covers the rationale of tissue engineering, the latest innovations, and the toxic potential of nanomaterials (NMs) used in the discipline of TERM.

8.2   Nanomaterials Used in TERM

The developing field of tissue engineering (TE) aims to regenerate damaged tissues by combining cells from the body with highly porous scaffold biomaterials, which act as templates for tissue regeneration, to guide the growth of new tissue. The aim of regenerative medicine (RM) is to replace degenerated or damaged tissues by combining stem cells, biomaterials, and physiochemical factors, such as growth factors. Applications and use of nanoscale technology such as nanofibers and nano-patterned surfaces have been used for directing cell behavior in the TERM field. Examples of different types of NPs which can be utilized for various applications in TERM include tissue targeting and imaging, bioactive agent delivery, modulating mechanical properties of scaffolds, and providing antimicrobial and antitumor properties.

Bone defects are commonly healed through the use of scaffolds which are comprised of osteoprogenitor cells, relevant biomaterials, and biochemical cues, such as growth factors. In bone tissue engineering (BTE), bone tissue scaffolds a 3D porous matrix consisting of cells, biocompatible materials, and essential biological and biochemical cues. The typical process for scaffold-based approaches for BTE can be seen in Fig. 8.1. The scaffold biomaterials have characteristic advantages like high penetration ability and high surface area with tunable surface properties, making them one of the widely preferred candidates in different fields of TERM. Classification

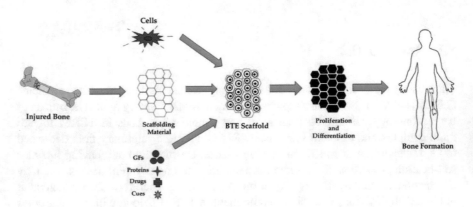

Fig. 8.1 Process for scaffold-based approaches for bone tissue engineering
https://www.mdpi.com/jfb/jfb-13-00001/article_deploy/html/images/jfb-13-00001-g001.png

and synthesis based on different criteria such as composition, structure, and manufacturing process have already been dealt with in Chap. 2.

A few commonly used NPs/NMs for TERM are summarized as under:

8.2.1 Metallic NPs

The unique physiochemical properties of metal NPs, such as antibacterial effects, shape memory phenomenon, low cytotoxicity, stimulation of the proliferation process, good mechanical and tensile strength, acceptable biocompatibility, significant osteogenic potential, and ability to regulate cell growth pathways, suggest that they can perform as novel types of scaffolds for BTE. Metallic NPs such as gold (Au) and Ag(Ag) NPs can be manufactured and modified by utilizing different functional groups that provide conjugation of antibodies, ligands, and drugs as delivery systems.

Ag NPs Ag NPs can also be described as a colloid of nanometer-sized particles of Ag and are one the most widely used metallic NPs in the biomedical field, mainly for their antimicrobial properties. Agions have been used for a long time for their antimicrobial properties toward a wide range of microorganisms. It has been shown that Agions are able to block the microbial respiratory chain system and precipitate bacterial cellular protein. NPs in the range of 1–10 nm can act differently against Gram- bacteria by (i) attaching to cell membrane affecting permeability and respiration, (ii) penetrating inside bacteria and damaging them, or (iii) releasing Agions (Chap. 3). Different scaffolds of Ag NPs can be produced in different formats including bulk materials, electrospun fibers, fibers mats, and nanofibrous or porous scaffolds. Delivery of Ag NPs not only has an antimicrobial effect but it also accelerates the rate of healing, and has the ability to regulate the cytokines associated with burn wound healing. Ag NPs also can show cytotoxic effects on cancer cells. A porous chitosan-alginate with biosynthesized Ag NPs also have cytotoxic effects against breast cancer cells. Therefore, it is very likely that these NPs have a potential for TERM.

Au NPs Colloidal Au NPs solutions present different properties compared to the bulk gold because of their optical property due to their unique interaction with light. Due to their strong affinity to gold, it is possible to conjugate various ligands including polypeptide sequences, antibodies, and proteins with various moieties such as phosphines, amines, and thiols including nuclear targeting of cancer cells. Therefore, Au NPs play a crucial role in the success of cancer treatment. Therefore, applying Au NPs prior to implantation can provide a safety measurement toolbox to minimize the recurrence of tumor through targeted delivery to cancer cells, and consequently increase the chance of effective implantation for various TERM applications.

8.2.2 Magnetic NPs

Magnetic nanoparticles (MNPs) are iron oxide (Fe3O4) NPs (or Fe_2O_3), which are widely studied in the biomedical field because of their low toxicity. Recently, superparamagnetic Fe3O4(SPIO)-Au core-shell NPs decorated with nerve growth factor (NGF) with low toxicity have been developed for neuron growth and differentiation. NGF functionalized NPs have provided higher neuronal growth, and orientation on PC-12 cells under dynamic magnetic fields utilizing rotation has been obtained compared to static magnetic fields. Magnetic NPs also have been used for controlling collagen fiber orientation dynamically and remotely in situ during the gelation period through an applied external magnetic field. Iron oxides also have the ability to pass the blood–brain barrier, which could be used for the conjugation of various peptides and growth factors to cure and regenerate brain tissue.

8.2.3 Ceramic NPs

Ceramic nanoparticles (CNPs) are primarily made up of oxides, carbides, phosphates, and carbonates of metals and metalloids such as calcium, titanium, silicon, etc. They have a wide range of applications due to a number of favorable properties, such as high heat resistance and chemical inertness. CNPs can be classified according to their tissue response as being (a) bioinert, (b) bioactive, or (c) resorbable ceramics and (d) magnetic NPs. In general, CNPs can be used in the production of nanoscale materials of various shape, size, and porosity.

Bioinert Nanoceramics Bioinert nanoceramics including TiO_2, ZnO are utilized for different medical applications as they show positive interactions with body tissues. TiO_2 NPs can be synthesized with different manufacturing processes including hydrothermal, solvothermal, sol-gel process, and emulsion precipitation methods. It is possible to manufacture uniformly distributed (in size) bioceramics in targeted size range. With the advancement of nanotechnology, TiO2 NPs, nanotubes, or nanoprobes labeled with fluorescent dye or magnetic resonance contrast agents have been successfully prepared for cell imaging through fluorescent analysis or magnetic resonance imaging (MRI). Similarly, Au NPs and Ag NPs, metal oxide NPs, and nanocomposite of chitosan/hydroxyapatite-zinc oxide (CTS/HAp-ZnO) supporting organically modified montmorillonite clay (OMMT) can be prepared for bone TE applications. Nanocomposite has shown strong antibacterial activities for both Gram+ and Gram- bacteria.

Bioactive Glass Ceramic NPs (n-BGC) n-BGC with SiO_2-CaO–P_2O_5-Na_2O core structures can be formed from various elements such as silicone, sodium, potassium, magnesium, phosphorous, oxygen, and calcium which can be absorbed by the cells. Antibacterial and angiogenic properties and excellent bioactivity of nBGC have made them a suitable candidate for dentin regeneration applications. The incorporation of boron-modified nBGC in the cellulose acetate/oxidized pullulan/ gelatin-based constructs has shown promising results for dentin regeneration through an increase in cellular viability.

Bioresorbable Nanoceramics Bioresorbable nanoceramics have a calcium phosphate (CaP)–based composition which includes a variety of materials such as hydroxyapatite (HAp), calcium aluminate, tricalcium phosphate, calcium phosphate dicalcium phosphate dehydrate, calcium carbonate ($CaCO_3$), calcium sulfate hemihydrate, octacalcium phosphate, and biphasic calcium phosphate. These materials have been applied in orthopedics, such as bone substitutes. HAp is commonly used in TERM applications. The combination of HAp with various forms of carriers such as electrospun fibers, porous scaffolds, and hydrogels can be used for the preparation of nanocomposite materials to modulate the desired cellular activities. Similarly, CaPs can be prepared with different types of polymers to produce nanocomposite materials. The advantages of these nanocomposites such as excellent mechanical characteristics could be utilized for bone tissue regeneration through enhancing scaffolds' performance.

8.2.4   Polymeric Nanoparticles

Low cytotoxicity of polymeric nanoparticles (PNPs), good biocompatibility, higher permeation and retention (EPR) effect, ability to deliver poorly soluble drugs and sustained release of them, and retaining bioactivity of bioactive agents from enzymatic degradation for TE applications make PNPs one of the fastest growing platforms to overcome obstacles in TERM. Most of the new PNP systems are designed to be sensitive to different physicochemical stimuli such as magnetic field, temperature, enzymes, pH, light, and reducing/oxidizing agents, which helps the delivery or targeting systems with high specificity and efficiency for TERM applications.

8.3   Scaffolds for TERM

Scaffolds are materials that have been engineered to cause desirable cellular interactions to contribute to the formation of new functional tissues for medical purposes. Cells are often "seeded" into these structures capable of supporting three-dimensional tissue formation. As such scaffolds represent important components for TE. However, researchers often encounter an enormous variety of choices when selecting scaffolds for TE. Apart from blood cells, most, if not all other, normal cells in human tissues are anchorage-dependent residing in a solid matrix called extracellular matrix (ECM). There are numerous types of ECM in human tissues, which usually have multiple components and tissue-specific composition. As for the functions of ECM in tissues, they can be generally classified into five categories. These

Table 8.1 Functions of extracellular matrix (ECM) in native tissues and of scaffolds in engineered tissues

Functions of ECM in native tissues	Analogous functions of scaffolds in engineered tissues	Architectural, biological, and mechanical features of scaffolds
1. Provides structural support for cells to reside	Provides structural support for exogenously applied cells to attach, grow, migrate, and differentiate in vitro and in vivo	Biomaterials with binding sites for cells; porous structure with interconnectivity for cell migration and for nutrients diffusion; temporary resistance to biodegradation upon implantation
2. Contributes to the mechanical properties of tissues	Provides the shape and mechanical stability to the tissue defect and gives rigidity and stiffness to engineered tissues	Biomaterials with sufficient mechanical properties filling up the void space of the defect and simulating that of the native tissue
3. Provides bioactive cues for cells to respond to their microenvironment	Interacts with cells actively to facilitate activities such as proliferation and differentiation	Biological cues such as cell-adhesive binding sites; physical cues such as surface topography
4. Acts as the reservoirs of growth factors and potentiates their actions	Serves as the delivery vehicle and reservoir for exogenously applied growth-stimulating factors	Microstructures and other matrix factors retaining bioactive agents in scaffold
5. Provides a flexible physical environment to allow remodeling in response to tissue dynamic processes such as wound healing	Provides a void volume for vascularization and new tissue formation during remodeling	Porous microstructures for nutrients and metabolites diffusion; matrix design with controllable degradation mechanisms and rates; biomaterials and their degraded products with acceptable tissue compatibility

supports include: structural support, mechanical support, providing bioactive cues for cells, acting as the reservoir of growth factors, and finally providing a flexible physical environment (Table 8.1).

8.4   Scaffolding Approaches

The application of hydrogel incorporated with metal NPs has become a new emerging research area in TERM. Disease, injury, and trauma often results in tissue damage and degeneration. By combining the hydrogel and NPs, the property enhancement of the materials can be achieved. This noble metal NPs have shown to have potential in TE applications. There are a few main approaches that have been adopted for the preparation of NP–hydrogel composites. Common examples of the preparation methods for biomedical applications (Fig. 8.2) include: (a) crosslinking of the

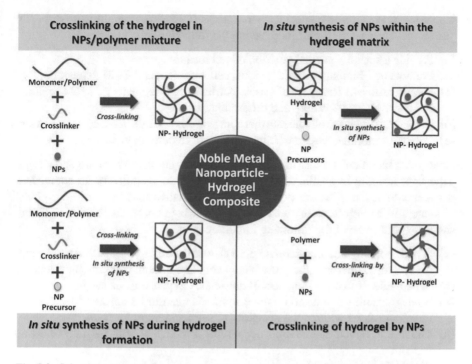

Fig. 8.2 Schematic diagram showing noble NP-hydrogel composite scaffolding approaches Bioengineering, https://www.mdpi.com/bioengineering/bioengineering-06-00017/article_deploy/html/images/bioengineering-06-00017-g001.png

hydrogel in NPs/polymer mixture, (b) in situ synthesis of NPs within the hydrogel matrix, (c) in situ synthesis of NPs during hydrogel formation, and (d) application of noble metal NP–hydrogel composites in TE. The potential of noble metal NPs and hydrogel in tissue regeneration have stimulated the high interest of researchers to further characterize the property of the hybrid of these materials. To address this, the studies on NP–hydrogel composites for the regeneration of tissues such as soft tissues, bone tissues, and cardiac tissues are being explored. The details of the synthesis of various scaffolds are out of the scope of this chapter.

8.5   Properties of Scaffolds/Matrices for TERM

As indicated previously, there are several scaffolding approaches in TERM. Each approach has its own pros and cons and preferred TE applications. In planning for TE for a complex tissue such as an intervertebral disc (IVD), these scaffolding approaches serve as important guidelines and can be used in combinations. Moreover, tissue-specific considerations in relation to the extent of injury, the unique structural functional relationship, multiple tissue composition, and interfaces in IVD deserve special attention. Scaffolds are three-dimensional (3D) porous solid biomaterials designed having the following properties:

(a) Provide a spatially correct position of cell location.
(b) Promote cell-biomaterial interactions, cell adhesion, and ECM deposition.
(c) Permit sufficient transport of gases, nutrients, and regulatory factors to allow cell survival, proliferation, and differentiation.
(d) Biodegrade at a controllable rate that approximates the rate of tissue regeneration.
(e) Provoke a minimal degree of inflammation or toxicity in vivo.

Apart from blood cells, most of the normal cells in human tissues are anchorage-dependent residing in a solid matrix (ECM). The best scaffold for an engineered tissue should be the ECM of the target tissue in its native state.

Generally, a scaffold, to serve as a suitable matrix to the reconstruction of tissue, should exhibit some of the following important features:

(a) A high porosity and an adequate pore size are necessary to facilitate cell seeding and diffusion throughout the whole structure of both cells and nutrients.
(b) Should allow effective transport of nutrients, oxygen, and waste.
(c) Biodegradability is essential, since scaffolds need to be absorbed by the surrounding tissues without the necessity of surgical removal.
(d) The rate at which degradation occurs has to coincide with the rate of tissue formation.
(e) Should be biocompatible.
(f) Should have adequate physical and mechanical strength. For example, the scaffolds in dentistry have an important distinction. In the bone, TERM requires a rigid scaffold that reproduces the size and architecture of the tissue to be rebuilt. In the pulpodentinal complex and in the periodontal apparatus of the TERM, due to the small size and difficulty to reach, the receiving site requires soft and injectable scaffolds. For this reason, the biomaterials used in scaffold formation can be classified according to the natural and synthetic sources or depending on the physical consistency, either rigid or soft.

8.6   Applications and Use in TERM

Novel cell sources, engineering materials, and tissue architecture techniques have helped to restore, maintain, improve, or replace biological tissues (Fig. 8.3). A number of therapies utilize NPs for the treatment of cancer, diabetes, allergy, infection, and inflammation. The surface conjugation and conducting properties of Au NPs, the antimicrobial properties of Ag and other metallic NPs and metal oxides, the fluorescence properties of quantum dots, and the unique electromechanical properties of carbon nanotubes (CNTs) have made them very useful in numerous TERM applications. In addition, magnetic NPs have been applied in the study of cell mechano-transduction, gene delivery, controlling cell patterning, and construction

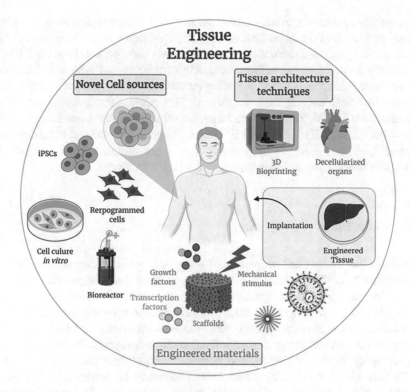

Fig. 8.3 Novel cell sources, engineering materials, and tissue architecture techniques have provided engineering tissues to restore, maintain, improve, or replace biological tissues
https://upload.wikimedia.org/wikipedia/commons/thumb/5/5b/Tissue_Engineering.png/330px-
Tissue_Engineering.png

of complex 3D tissues. The use of the right type of NPs can significantly enhance the biological, mechanical and electrical properties of scaffolds as well as can serve various functions in TERM.

8.6.1   Biological Properties

Au NPs and *titanium dioxide* (TiO$_2$) NPs have been used to enhance cell proliferation rates for bone and cardiac tissue regeneration, respectively. Au NPs have shown superior biocompatibility and the ability for surface modification, which has resulted in interesting biomedical applications. These NPs can be described as a colloid of nanometer-sized particles of Au. Colloidal Au solutions present different properties compared to bulk Au because of their optical property that provides unique interaction with light. On Au surface it is possible to conjugate various ligands including polypeptide sequences, antibodies, and proteins with various moieties such as phosphines, amines, and thiols because of their strong affinity to Au. In bone, Au NPs promote osteogenic differentiation of an osteoblast precursor cell line, MC3T3-E1. In addition, these NPs also influence osteoclast (or bone resorbing cell) formation from hematopoietic cells while providing protective effects on mitochondrial dysfunction in osteoblastic cells. Therefore, Au NPs present themselves as excellent candidates for TERM. They seem to be perfect candidates to replace bone morphogenetic proteins (BMPs). Although BMPs have beneficial effects on bone regeneration and repair, they also have many disadvantages such as high cost and susceptibility to result in unwanted bone formation and local inflammatory reactions.

8.6.2   Mechanical Properties

Change in the mechanical properties of some NPs leads to superior mechanical properties for TERM applications compared to scaffolds without NP reinforcements. For example, NP-embedded nanocomposite polymers both in the form of hydrogels and electrospun fibers exhibited an enhanced response than simple NPs. Likewise, a TiO_2-embedded biodegradable patch showed a higher tensile strength in reinforcing the scar after myocardial infarction. Hydrogel microfibers with poly(*N*-isopropylacrylamide) (PNIPAm) and MNPs increased their mechanical strength. CNTs have also been used to enhance the mechanical properties of polymers for TERM applications. CNTs reinforce the polymers especially due to their remarkable mechanical properties, tensile strength, and fiber-like structure.

8.6.3   Electrical Properties

Nanoparticles have also been used to enhance the electrical properties of scaffolds, which can be highly beneficial in cardiac TERM. Au NP-based electrospun fibrous scaffolds are excellent for cardiac tissue regeneration. Au NPs are deposited on the surface of the gelatin and PCL–gelatin fibers, creating nanocomposites with a nominal Au thickness. Cardiac tissues engineered within these Au NP scaffolds can be used to improve the function of the infarcted heart. Likewise, Au nanowires have also been used as conductive materials alongside scaffolds to enhance the electrical coupling between the cells. With time, cardiac muscle cells start growing within the 3D porous scaffolds and result in synapse formation. The use of CNTs in polymer composites can significantly improve conductivity to promote cardiomyocyte functions, therefore conductive nanomaterials have a promising future in cardiac applications.

8.6.4 Antibacterial Properties

Some metal oxides, particularly Ag NPs, have shown great antimicrobial effects, as well as wound-healing capabilities. Likewise, Ag-containing poly(3-hydroxybutyrate-co-3-hydroxyvalerate) (PHBV) nanofibrous scaffolds had high antibacterial properties, and they exhibit excellent in vitro cell compatibility. This shows that PHBV nanofibrous scaffolds containing Ag NPs have prospects to be used in joint arthroplasty. Biocomposite scaffolds containing nano Ag could regulate bacterial infection during reconstructive bone surgery. Thus, the presence of Ag NPs in the scaffolds acts as an affixed coating for protection against infection, sepsis, and malfunctioning of implants. Specifically, Fe_3O_4 NPs also have much promise in killing post-biofilm formation in bacteria (especially when functionalized with sugars such as fructose and sucrose) and can penetrate biofilms, whereas antibiotics cannot (magnetic field disrupts and kills bacteria). This has enormous consequences in TERM since currently if a biomaterial becomes infected, it needs to be removed and adjacent tissue cleaned. Strategies that do not rely on implant removal and cleaning can have a bright future in TERM.

8.6.5 Gene Delivery

Gene delivery has been used for regenerative medicine applications to create or restore normal function at the cell and tissue levels. For effective gene therapy applications, it is vital to build up a suitable vector system with a high gene transfection efficiency, low cytotoxicity, and high specificity to unhealthy cells. Gene delivery is classified into two categories: nonviral and viral.

The advantages of nonviral methods are their simplicity and the absence of an immune response, while the disadvantage is low efficiency due to low transfection rate. The promise for gene delivery has been seen through the use of NPs and self-assembled NMs. Magnetofection is a new method for gene delivery in which gene transfection is accomplished using magnetic NPs. To achieve magnetofection using plasmid DNA, cationic lipids or polymers with complexes of DNA interact with magnetic beads and then through a magnetic force are attracted onto target cells so that they can accumulate on the surface. For example, Fe_3O_4 magnetic particles can attach to the gene and improve transfection efficiency. These particles are distributed within a polymer matrix or internalized in a polymer or metallic case, which binds DNA through charge contact. CNTs, which have shown numerous applications and can be synthesized through several different approaches, have also shown remarkable potential as nonviral gene delivery agents. A technique called nanotube spearing can be used to prepare nickel-embedded, magnetic nanotubes where DNA is attached.

Viral transduction methods have also been used in TERM applications; however, one problem associated with this method is the difficulty in preparing viral vectors with a high titer.

8.6.6 Mechano-Transduction

In addition to various bioactive molecules and growth factors that regulate cell functions in the human body, mechanical forces play a major role in determining cell functions by affecting mechano-transduction pathways. Numerous approaches such as the introduction of shear stress by bioreactors and stiffness of patterned substrates to mechanically control cell functions have been used. However, magnetic NPs have proven to be superior to all these methods since they can be controlled remotely, spatially, and temporally through a magnetic field. In this process, first, the MNPs are coated with a certain targeting antibody. Once the magnetic field is applied, the cells are clustered in the direction of the magnetic field. Based on the antibody used, receptor-mediated cell function is affected.

8.6.7 Magnetic Cell Patterning

TERM requires the fabrication of tissue architecture similar to in vivo conditions. To get cell adhesion within a specified design pattern magnetic cell-patterning technique is used. A cell-patterning technique has been developed using magnetite cationic liposome (MCL), where a magnet with a magnetic field concentrator is laid under a cell culture surface. Various cell patterns can be successfully fabricated using this technique by manipulating the line patterns of the magnetic field concentrators. When human umbilical vein endothelial cells are used, the cells connect and form capillary-like structures with patterned in a line.

8.7 Constructing 3D Tissues

Three-dimensional bioprinting is a rapidly growing technology that has been widely used in TE, disease studies, and drug screening. It provides the unprecedented capacity of depositing various types of biomaterials, cells, and biomolecules in a layer-by-layer fashion, with a precisely controlled spatial distribution. There are three categories of 3D bioprinting strategies: inkjet-based printing (IBP), extrusion-based printing (EBP), and light-based printing (LBP).

Bioinks are formed by combining cells and various biocompatible materials, which are subsequently printed in specific shapes to generate tissue-like, 3D structures. These bioinks mimic the ECM environment, support cell adhesion, proliferation, and differentiation after printing. In contrast to traditional 3D printing materials, bioinks must have:

- Print temperatures that do not exceed physiological temperatures.
- Mild cross-linking or gelation conditions.
- Bioactive components that are non-toxic and can be modified by the cells after printing.

Three-dimensional bioprinting allows for the spatially controlled placement of cells in a defined 3D microenvironment. Due to the high degree of control on structure and composition, 3D bioprinting has the potential to solve many critical unmet needs in medical research, including applications in cosmetics testing, drug discovery, regenerative medicine, and functional organ replacement. Personalized models of disease can be created using patient-derived stem cells, such as induced pluripotent stem cells (iPS cells) or mesenchymal stem cells. Depending on the application, a range of materials, methods, and cells can be used to yield the desired tissue construct. 3D bioprinting allows for the spatially controlled placement of cells in a defined 3D microenvironment. This technology is expected to address the organ-shortage issue in the future. The combination of stem cell technology and 3D bioprinting is expected to allow the construction of better-functioning tissues/organs and organs-on-chips. For the longevity and functionality of printed architectures, vascularization and innervation need to be further investigated.

8.8   Bioactive Agents/Molecules

Bioactive agents act as a scaffolding frame to deliver cells to the appropriate site, define a space for tissue development, and direct the shape and size of the engineered tissue. Agents including proteins and small molecules involved in TERM play an important role in controlling the microenvironment in vivo. Chemotactic signals from bioactive molecules are responsible to regulate host cell migration, proliferation, and differentiation and allow cells to interact via specific receptors for chemical recognition with their surrounding microenvironment. Moreover, an anatomic destination is identified according to certain concentration gradients of chemicals produced at injured sites within the microenvironment. Thus incorporation of a suitable bioactive molecule through the design of a tissue-engineered scaffold can promote tissue regeneration by stimulating the transplanted cells or adjacent host cells. The mode of release is especially relevant when the bioactive agent is a growth factor (GF) because the dose and the spatiotemporal release of such agents at the site of injury are crucial to achieve a successful outcome. Strategies that combine scaffolds and drug delivery systems have the potential to provide more effective tissue regeneration relative to current therapies. NPs can protect the bioactive agent, control its profile, decrease the occurrence and severity of side effects, and deliver the bioactive agent to the target cells maximizing its effect. Scaffolds containing NPs loaded with bioactive agents can be used for their local delivery, enabling site-specific pharmacological effects such as the induction of cell proliferation and differentiation, and, consequently, neo-tissue formation. Recently, plasma protein–based NPs have gained attention due to their high bioavailability, non-toxicity, biodegradability, ease of manipulation, long in vivo half-lives, and long shelf lives. There are more than 100,000 proteins in human plasma, but just a couple of these proteins have been used in TERM as a nanocarrier platform for imaging, drug delivery, and tissue regeneration. High density lipoproteins (HDL) NPs are among candidates for enhancing photodynamic therapy applications through presenting excellent tumor targeting and internalization capacity. Fibrin is another plasma protein that has been used for encapsulation of vascular endothelial growth factor (VEGF) for promoting angiogenesis for wound-healing applications. NPs from albumin, as the most abundant plasma protein, are being used for bone regeneration through a sustained release of bone morphogenetic protein-2 (BMP-2). Bone tissue–related diseases such as a tumor or trauma generally are treated with bone grafts and substitutes. Nowadays TERM has provided an alternative approach for bone tissue regeneration by offering a variety in forms of 3D scaffolds. Bone scaffolds containing stem cells have the advantage of controlling the cellular activity such as differentiation, if appropriate bioactive agents such as drugs (e.g., dexamethasone) or growth factors [e.g., bone morphogenetic proteins (BMPs)] are incorporated in them.

Ascorbic acid is an important water-soluble bioactive molecule in the bone formation process. It is also called vitamin C, acts as a cofactor for the key enzymes involved in collagen biosynthesis, and demonstrates a major function in stabilizing the helical structure of collagen. Bone tissue–related diseases such as a tumor or trauma generally are treated with bone grafts and substitutes. Nowadays TERM has provided an alternative approach for bone tissue regeneration by offering a variety in forms of 3D scaffolds. Natural biodegradable multi-channeled scaffolds composed of ordered electrospun nanofibers with neurotrophic gradient have been designed to control axon outgrowth. Likewise, various NP-based dressings have been developed for delivering bioactive agents with spatiotemporal control for enhancing the wound-healing process.

8.9 Imaging and Contrast Agents

Imaging strategies, in conjunction with exogenous contrast agents, can help in assessing in vivo therapeutic progress of TE. Proper use of these monitoring/imaging and regenerative agents (MIRAs) can help increase TERM therapy successes and allow for clinical translation. MIRA research is still in its beginning stages with much of the current research being focused on imaging or therapeutic applications, separately. Advancing MIRA research will have numerous impacts on achieving clinical translations of TERM therapies.

8.10 Biocompatibility

Materials obtained using nanotechnology and currently used in TERM include a wide range of products. In order to overcome the challenges of high organ demand and biocompatibility issues, scientists in the field of TREM are working on the use of scaffolds as an alternative to transplantation. For complex scaffolds of different compositions and structures, the scaffolds are being developed to mimic the ECM, act as structural support, and define the potential space for new tissue development as well as enhance cell attachment, proliferation, and differentiation. Tissue-engineered products (TEPs) containing either cells or growth factors or both cells and growth factors may be used as an alternative to the autografts taken directly from the bone of the patients. Nevertheless, the use of TEPs needs much more understanding of biointeractions between biomaterials and eukaryotic cells. Despite the possibility of the use of in vitro cellular models for the initial evaluation of the host response to the implanted biomaterial, it is observed that most researchers use cell cultures only for the evaluation of cytotoxicity and cell proliferation on the biomaterial surface, and then they proceed to animal models and in vivo testing of bone implants without fully utilizing the scientific potential of in vitro models. However, nanostructured hydroxyapatite (nano-HA) and nano CaP have received considerable attention. The nano-HA has shown excellent biological performances compared to conventional HA because it has better biocompatibility and bioactivity in respect of bone components (probably as a result of its similarity with the chemical component and mineral structure of bone tissue). Moreover, due to their small size and large specific surface, nano-HAs may not only promote ion exchange within a physiological environment but also increase protein absorption and cellular response, especially if stressed by physical means.

8.11 Nanotoxicity of NMs

The main risks in TE are tumorigenity, graft rejection, immunogenity, and cell migration. Secondly, the toxicity of NPs is highly dose- and exposure-dependent. In many applications, the NPs are used below their threshold concentrations at which they are considered not harmful. However, bioaccumulation of NPs inside the body over a large period of time is well known. Thus, any NP used in the human body has the potential to accumulate over a long period of time to reach a concentration that can cause toxicity to cells, cancers, or harmful effects on reproductive systems and fetuses before their birth. In addition, even though numerous products containing NPs/NMs are already on the market, there are still some scientific and methodological gaps in the knowledge of specific hazards of NMs. As has been indicated previously (Chap. 5), currently there are no international standards yet for nano-specific risk assessments, including specific data requirements and testing strategies. The risk assessments of NMs are laborious and costly. Currently, manufacturers are committed to assess the safety of their NP-based products and to implement the necessary safety measures (self-supervision). To date, the regulatory tools are not nano-specific; for example, the data requirements for notification of chemicals, criteria for classification, and labeling requirements for safety data sheets are still not widely available. Thus, there is a need for precautionary measures for the applications of NPs wherever there is a possibility of chronic bioaccumulation.

8.12 Conclusion

Nanoparticles exhibit superior biocompatibility and well-established strategies for surface modification, which have made them highly effective in numerous biomedical applications. The inability to deliver bioactive agents locally in a transient but sustained manner is one of the challenges in the development of bio-functionalized scaffolds for TERM. In vivo maturation of engineered tissues requires a well-synchronized series of events involving the host immune system, circulatory system, and cellular components of the implanted material. The physicochemical properties of the scaffolds and the presence/immobilization of bioactive agents within the scaffolds can help in achieving precise control over maturation. NPs have been shown to develop optical, electrical, and electrochemical biosensors for molecules, proteins, and DNA detection with highly accurate results. By optimizing these functions, these sensors could have a great influence on medical use. Although NPs show a promising future in TERM applications, there is still a lack of in vivo experimentation that needs to be done in order to verify the wide variety of successful results from in vitro studies. However, the numerous existing applications in the literature reiterate the great potential NPs can have on TERM through bioinks, imaging, and contrast biosensors. At the same time, it is important to look into the risks associated with TERM which are tumorigenity, graft rejection, immunogenity, and cell migration. Therefore, our aim is to understand the risks, how to minimize them and, especially, how to predict and prevent them.

Further Reading

Fang YL, Chen XG, Godbey WT. Gene delivery in tissue engineering and regenerative medicine – a review. J Biomed Mater Res B Appl Biomater. 2015 Nov;103(8):1679–99. https://doi.org/10.1002/jbm.b.33354.

Fathi-Achachelouei M, Knopf-Marques H, Ribeiro da Silva CE, Barthès J, Bat E, Tezcaner A, Vrana NE. Use of nanoparticles in tissue engineering and regenerative medicine. Front Bioeng Biotechnol. 2019;7:113. https://doi.org/10.3389/fbioe.2019.00113.

Gupta PK. Chapter 15: Toxic effects of nanoparticles. In: Toxicology: resource for self study questions. 2nd ed. Seattle: Kinder Direct Publications; 2020a.

Gupta PK. Chapter 14: Toxicology of nanoparticles. In: Problem solving questions in toxicology – a study guide for the board and other examinations. 1st ed. Cham: Springer; 2020b.

Gupta PK. Toxic effects of nanoparticles. In: Brain storming questions in toxicology. 1st ed. Boca Raton: Taylor & Francis Group, LLC, CRC Press; 2020c. p. 297–300.

Gupta PK. Fundamentals of Nanotoxicology. 1st ed. New York: Elsevier Inc.; 2022.

Hasan A, Morshed M, Memic A, Hassan S, Webster TJ, Marei HE. Nanoparticles in tissue engineering: applications, challenges and prospects. Int J Nanomed. 2018;13:5637–55. Published 2018 Sept 24. https://doi.org/10.2147/IJN.S153758.

Khanna P, Ong C, Bay BH, Baeg GH. Nanotoxicity: an interplay of oxidative stress, inflammation and cell death. Nanomaterials. 2015;5:1163–80. https://doi.org/10.3390/nano5031163.

Rychter M, Baranowska-Korczyc A, Luleka J. Progress and perspectives in bioactive agent delivery via electrospun vascular grafts. RSC Adv. 2017;7:32164.

Tan H-L, Teow S-Y, Pushpamalar J. Application of metal nanoparticle–hydrogel composites in tissue regeneration. Bioengineering. 2019;6(1):17. https://doi.org/10.3390/bioengineering6010017.

Walmsley GG, McArdle A, Tevlin R, Momeni A, Atashroo D, Hu MS, Feroze AH, Wong VW, Lorenz PH, Longaker MT, Wan DC. Nanotechnology in bone tissue engineering. Nanomedicine. 2015;11(5):1253–63. https://doi.org/10.1016/j.nano.2015.02.013.

Zhang B, Gao L, Ma L, Luo Y, Yang H, Cui Z. 3D bioprinting: a novel avenue for manufacturing tissues and organs—review. Res 3D Bioprint. 2019;5(4):777–94. https://doi.org/10.1016/j.eng.2019.03.009.

Chapter 9
Nanotechnology in Cancer Therapy

Abstract As cancerous cells grow and multiply, they form a mass of cancerous tissue. Cancer is a leading cause of morbidity and mortality worldwide. Despite efforts to mitigate risk factors in recent decades, the prevalence of cancer is continuing to increase. Nanotechnology has been extensively studied and exploited for cancer treatment as NPs play a significant role as a drug delivery system. Compared to conventional drugs, NP-based drug delivery has specific advantages, such as improved stability and biocompatibility, enhanced permeability and retention effect, and precise targeting. The application and development of hybrid NPs, which incorporate the combined properties of different NPs, have led this type of drug-carrier system to the next level. This chapter discusses the use of therapeutic systems using external stimuli and self-therapeutic nanomaterials (NMs) along with their mechanisms and possible toxicological implications and health hazards associated with their use to target and eradicate cancer cells. However, there is an urgent need for researchers to develop NMs and successfully transition them from the bench to the clinics.

Keywords Drug delivery · Cancer therapy · Engineered NMs · Next-generation · Nanotoxicity · Cancer · Self-therapeutic NMs · Target delivery · Prodrug · Drug · NPs · Toxicity

9.1 Introduction

Cancer is a leading cause of morbidity and mortality worldwide. Despite efforts to mitigate risk factors in recent decades, the prevalence of cancer is continuing to increase. Nanomedicine represents an innovative field with immense potential for improving cancer treatment, having ushered in several established drug delivery platforms. Currently, the field of nanomedicine is generating a new wave of nanoscale drug delivery strategies, embracing trends that involve the

© The Author(s), under exclusive license to Springer Nature Switzerland AG 2023   143
P K Gupta, *Nanotoxicology in Nanobiomedicine*,
https://doi.org/10.1007/978-3-031-24287-8_9

functionalization of these constructs with moieties that enhance site-specific delivery and tailored release. This chapter briefly summarizes several advancements in established nanoparticle (NP) technologies such as liposomes, polymer micelles, and dendrimers regarding tumor targeting and controlled release strategies and the toxic paradigms of chemotherapeutic drugs and NPs used in cancer therapy.

9.2   Overview of Cancer

A cancer is an abnormal growth of cells (usually derived from a single abnormal cell). The cells have lost normal control mechanisms and thus are able to multiply continuously, invade nearby tissues, migrate to distant parts of the body, and promote the growth of new blood vessels from which the cells derive nutrients. Cancerous (malignant) cells can develop from any tissue within the body. As cancerous cells grow and multiply, they form a mass of cancerous tissue—called a tumor—that invades and destroys normal adjacent tissues. The term "tumor" refers to an abnormal growth or mass. Tumors can be cancerous or noncancerous. Cancerous cells from the primary (initial) site can spread throughout the body (metastasize).

The development and spread of cancer involve three stages such as initiation, promotion, and spread. The first step in cancer development is initiation, in which a change in a cell's genetic material (a mutation) primes the cell to become cancerous. The change in the cell's genetic material may occur spontaneously or be brought on by an agent that causes cancer (a carcinogen).

The second step in the development of cancer is promotion. Agents that cause promotion, or promoters, may be substances in the environment or even some drugs such as sex hormones (e.g., testosterone used to improve sex drive and energy in older men). Unlike carcinogens, promoters do not cause cancer by themselves. Instead, promoters allow a cell that has undergone initiation to become cancerous. Promotion has no effect on cells that have not undergone initiation. Some carcinogens are sufficiently powerful to be able to cause cancer without the need for promotion. For example, ionizing radiation (which is used in x-rays and is produced in nuclear power plants and atomic bomb explosions) can cause various cancers, particularly sarcomas, leukemia, thyroid cancer, and breast cancer.

Cancer can grow directly into (invade) surrounding tissue or spread to tissues or organs, nearby or distant. Cancer can spread through the lymphatic system. This type of spread is typical of carcinomas. For example, breast cancer usually spreads first to the nearby lymph nodes in the armpit, and only later it spreads to distant sites. Cancer can also spread via the bloodstream. This type of spread is typical of sarcomas.

As a cancer grows, nutrients are provided by direct diffusion from the circulation. Local growth is facilitated by enzymes (e.g., proteases) that destroy adjacent tissues. As the cancer volume increases, the cancer may release angiogenesis factors, such as vascular endothelial growth factor (VEGF), which promotes the formation of new blood vessels that are required for further growth.

In brief, tumors or cancer cells are abnormal in appearance and other characteristics, having undergone one or more of the following alterations: (1) hypertrophy, or an increase in the size of individual cells; this feature is occasionally encountered in tumors but occurs commonly in other conditions; (2) hyperplasia, or an increase in the number of cells within a given zone; in some instances, it may constitute the only criterion of tumor formation; (3) anaplasia, or a regression of the physical characteristics of a cell toward a more primitive or undifferentiated type; this is an almost constant feature of malignant tumors, though it occurs in other instances both in health and in disease. As a tumor grows larger, it invades the healthy tissues nearby. In cancer, metastasis involves three primary processes: invasion/intravasation, extravasation, and angiogenesis (Fig. 9.1). Tumor cells invade through the extracellular matrix (ECM) or vascular endothelium into the circulation, then

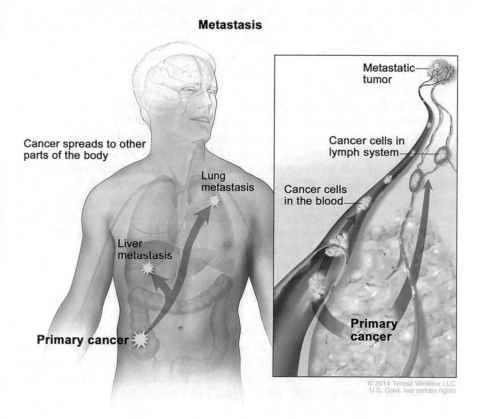

Fig. 9.1 In metastasis, cancer cells break away from where they first formed and form new tumors in other parts of the body (©2014 Terese Winslow LLC. U.S. Govt. has certain rights)
https://www.cancer.gov/sites/g/files/xnrzdm211/files/styles/cgov_article/public/cgov_contextual_image/900/300/files/metastasis-enlarge.jpg?itok=CBtbX_qG

extravasate and colonize into other organs, and form micrometastases. Because of the vascular system, the endothelial wall represents a natural barrier to the migrating tumor cells that need to be breached during the intra-/extravasation process. Intravasation is a process where tumor cells travel beyond the endothelial basement membrane of blood vessels to enter the circulation. Extravasation is a process where tumor cells travel across the vessel wall to leave the circulation and enter into a metastatic site/organ. It is the last and rate-limiting step before secondary tumor formation, which plays a vital role in cancer development and metastasis.

9.3   NPs/NMs Used in Cancer Therapy

Recently, NPs/NMs have attracted much attention from scientists interested in cancer therapy because of their versatile physical and chemical properties. These relatively small particles can also be functionalized with ligands, nucleic acids, peptides, or antibodies that bind to specific target molecules. Because of certain intrinsic properties of NPs/NMs (reaction with oxygen species, heat, and hazardous gas producers) and biocompatibilities, they are quite interesting to use directly for therapeutic purposes. Most of the organic NPs {liposomes, micelles, exosomes, lipids, PEGylated polylactide (*PLA*) and poly lactic-co-glycolic acid (*PLGA*)}, inorganic NPs (gold, silver, silica, iron, graphene, carbon quantum dots), and composites ({metal−organic frameworks (MOF)}, transition metals dichalcogenide (TMD)) have been designed as carriers in drug-delivery systems for use in external stimuli-based systems such as photodynamic therapy (PDT), photothermal therapy (PTT), magnetic therapy, and boron neutron capturing therapy (BNCP) and many reports have focused on the use of NMs as carriers of therapeutic compounds. These therapeutic systems have been extended by introducing external stimuli (e.g., light, magnetic waves, and heat) to improve drug release at the tumor sites. Such nanosystems for therapeutic applications is limited to drug delivery and theranostic agents using external stimuli-responsive systems (Fig. 9.2).

Unfortunately, the use of NMs as carriers of therapeutic compounds has some flaws, which preclude these therapeutic systems from clinical applications. The major challenges are a low drug loading efficiency, low solubility in an aqueous media, poor ability to cross in vivo barriers and penetrate inside the tumor (less than 1% reach the tumor), problematic physical and chemical interactions of hydrophobic therapeutic compounds with NMs, in vivo instability, a suboptimal biodistribution, low tumor targeting ability, and a suboptimal drug release profile. To minimize all of these issues, new NMs are being developed that can reduce the growth of aggressive tumors through their self-therapeutic properties.

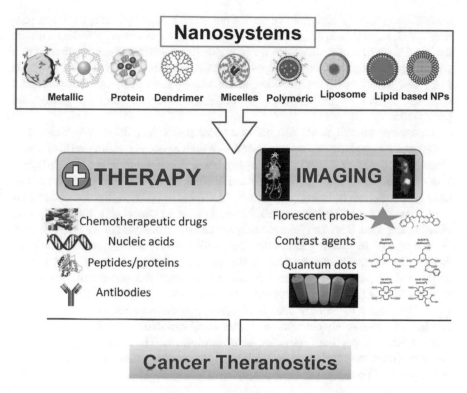

Fig. 9.2 Nanosystems for cancer therapy: therapy and imaging (polymeric, lipid based, and metallic NMs for cancer theranostics)
https://www.frontiersin.org/files/Articles/488377/fphar-10-01264-HTML/image_m/fphar-10-01264-g001.jpg

9.3.1 External Stimuli-Responsive Systems

Drug Delivery

Targeted drug delivery nanosystems include polymeric micelles, polyelectrolyte complex micelles, liposomes, dendrimers, nanoemulsions, and NPs (see also Chaps 2 and 4). They have a unique core/shell nanostructure where the hydrophobic segments of the amphiphilic copolymers form an inner core surrounded by a layer of hydrophilic segments. These polymer assemblies have attracted much attention as useful drug carriers for cancer therapy because they are able to incorporate water-insoluble anti-cancer drugs, such as paclitaxel and camptothecin, within their hydrophobic core. In addition, the hydrophilic shell layer of these micelles can protect the incorporated drugs from degradation by enzymes, avoid non-selective uptake by macrophages distributed in the body, and hence allow for targeting cancerous tissues via the enhanced permeability and retention (EPR) effect. Polyelectrolyte complex micelles are usually produced by electrostatic interactions between anionic macro-molecules and the di-block copolymers composed of cationic segments and hydrophilic segments. In contrast to the polymeric micelles, the polyelectrolyte complex micelles can be extensively applied for intravenous and intracellular delivery of various bioactive macromolecules including peptides, proteins, and nucleic acids. Moreover, polyelectrolyte complex micelles can be designed with different structures and compositions, making them applicable to a wide range of biopharmaceutical applications. A number of other targeting ligands (e.g., folic acid, transferrin, peptides, or antibodies) have been covalently coupled onto the surface of the micelles for achieving effective tumor targeting and penetration.

Liposomes provide many distinctive advantages as drug delivery vehicles such as carrying hydrophilic therapeutic agents into their aqueous interior and also water-insoluble anti-cancer drugs into the hydrophobic domain within the lipid bilayers. They also have relatively high drug loading capacity so that tens of thousands of drug molecules can be entrapped in their structures. In addition, liposomes are generally considered to be biocompatible and known to cause very little antigenic, allergic, and toxic reactions because they are usually composed of naturally derived phospholipids. In recent years, nanoscale delivery vehicles for small interfering RNA (siRNA) have been also developed as effective therapeutic approaches to treat cancer. siRNA has been incorporated into various types of nanoparticulate formulations including liposomes, polyelectrolyte complexes, polymeric NPs, and inorganic NPs. The most widely used methodology to deliver nucleic acid drugs is to condense the negatively charged molecules using cationic materials, made of polymers, lipids, and peptides, into nano-sized complexes. The nano-complexed particles containing the siRNA are then delivered to the target cells more efficiently due to the physical protection of the RNA to the host's systems.

Theragnostic Agents
Nanoparticles have considerably improved the diagnostics and therapeutics of various cancers due to their small size, ease of functionalization, enhanced drug loading (due to large surface-to-volume ratio), effortless penetration abilities, and improved retention inside target tissue. Apart from that, excellent biocompatibility, biodegradability, and multifunctional applications, including bio-imaging, bio-sensing, diagnostics, and therapeutics, have increased the potential use of these NMs for various biomedical applications. Usually, the NP surface is modified using a ligand in order to target specific tumor cell molecules. As more ligands are attached to the NP surface, there are more chances to bind the target cell. The amount of signaling groups influences the sensitivity of the detection method. Some NPs have innate optical properties like quantum dots (QDs) and metallic NPs due to surface plasmon resonance. QDs NPs labeled with 18F-fluoropropionate and functionalized with Arginylglycylaspartic acid (RGD) peptides demonstrate proper optical characteristics for PET imaging of prostate cancer because of high spatial resolution and high sensitivity for CT imaging. By functionalization with chitosan polymers, they can be used for colorectal adenocarcinoma imaging. Also, they can be conjugated with antibodies for lymph nodes and metastases imaging in squamous cell carcinoma, and head and neck cancer. Moreover, gold NPs radiolabeled with Indium-111 (^{111}In) and Iodine-125 (^{125}I) can be used in SPECT imaging of epidermoid carcinoma.

Iron oxide NPs are widely used in MRI imaging because they can improve and enhance the contrast. In glioblastoma, iron oxide NPs functionalized with peptides and polymers accumulate within the tumor microenvironment by forming self-assembly structures. In addition, polymeric materials such as mesoporous silica NPs carry tumor-targeting properties and are proposed for PET imaging in breast cancer. Besides this, they perform drug delivery applications. In addition, other NP formulations (nanoliposomes, micelles, polymersomes, dendrimers, and aptamers) that need to be functionalized with specific contrast agents and fluorophores can be effectively used for imaging. The advantages to implement NPs such as molecular imaging tools are biocompatibility and biodegradability, encapsulation properties, water solubility in some cases, and targeting ligands accessibility.

Advantages of NMs as Drug Carriers The use of nanocarriers for drug delivery offers many advantages:

(i) Circumvents the problems of solubility and stability of anticancer drugs.
(ii) Prevents the drug from degradation from proteases and other enzymes and increases the half-life of the drug in the systemic circulation.
(iii) Improves drug distribution and targeting.
(iv) Helps in the sustained release of the drug by targeting the cancer sites.
(v) Helps in the delivery of multiple drugs and, therefore, in reducing drug resistance.

Another advantage of these NPs is their application to capture cancer biomarkers, such as cancer-associated proteins, circulating tumor DNA, circulating tumor cells, and exosomes. An essential advantage of applying NPs for cancer detection lies in their large surface area to volume ratio relative to bulk materials. Due to this property, NP surfaces can be densely covered with antibodies, small molecules, peptides, aptamers, and other moieties. These moieties can bind and recognize specific cancer molecules. By presenting various binding ligands to cancer cells, multivalent effects can be achieved, which can improve the specificity and sensitivity of an assay.

9.3.2 Self-Therapeutic NMs

Self-therapeutic NMs are successfully used to target and eradicate cancer cells. Many reports have focused on the use of NMs as carriers of therapeutic compounds. These relatively small particles can also be functionalized with ligands, nucleic acids, peptides, or antibodies that bind to specific target molecules. Because of certain intrinsic properties of NMs (reaction with oxygen species, heat, and hazardous gas producers) and biocompatibilities, they are quite interesting to use directly for therapeutic purposes. For example, boron-based compounds (metalloid boron as an enzyme inhibitor) are directly used for the growth inhibition of different cancer cells. Boric acid (BA), the predominant form of boron in plasma, shows boron-mediated anticancer mechanisms in prostate cancer by reducing intracellular calcium signals and calcium storage; decreasing enzymatic activities (serine prote-ase, NAD-dehydrogenases, etc.); and finally inhibiting cancer cell proliferation. Another compound, magnesium silicide NPs (MS NPs) could be used as a deoxy-genation agent for cancer therapy. Likewise, self-therapeutic organic NMs such as peptide-based NMs are used in medicine for therapeutic applications. The major advantage of peptide-based therapeutic systems is that they could target cancer cells with lower toxicity to normal tissues.

9.3.3   Other Self-Therapeutic NMs

Recently it has been reported that biodegradation of graphene-based NMs in blood plasma have in vitro and in vivo antitumor abilities. The bio-transformed materials exhibit high efficiencies of drug delivery and more pronounced targeting and tumor-killing abilities. Likewise, sodium chloride NPs (SCNPs) synthesized from sodium oleate, molybdenum chloride, and oleylamine as a surfactant lead to asymmetric ionic gradients that are involved in the regulation and control of cell function. Reducing and increasing extracellular sodium and chloride concentrations can cause cytoskeleton destruction, cell cycle arrest, and cell lysis.

9.4   Benefits of Nanotechnology

As discussed earlier, NPs/NMs are fabricated using organic, inorganic, lipid, and protein compounds typically in the range of 1–100 nm. They are similar in size to large biological molecules ("biomolecules") such as enzymes and receptors. For example, hemoglobin, the molecule that carries oxygen in red blood cells, is approx-imately 5 nm in diameter. Nanoscale devices smaller than 50 nm can easily enter most cells, while those smaller than 20 nm can move out of blood vessels as they circulate through the body. Because of their small size, nanoscale devices can read-ily interact with biomolecules on both the surface and inside cells. By gaining access to so many areas of the body, they have the potential for treatment and ther-apy, and for detection and diagnosis.

9.4.1   Treatment and Therapy

During the past 70 years, the number of cancer deaths has continued to rise, as com-pared to the slight increase in the number of people died of other diseases such as heart diseases, cerebrovascular diseases, and pneumonia. During this period, radio-therapy and chemotherapy were the principal treatment modalities aimed at eradi-cating solid tumors that are located deep inside the body. However, these methods have suffered from their non-specific mode of action, which not only killed cancer cells but also harmed normal cells at the same time. For example, the most common chemotherapeutic agents such as paclitaxel and doxorubicin exhibit anti-cancer effects by inducing apoptotic death of rapidly dividing cells, but they can also kill several types of normal cells that divide rapidly in ordinary circumstances. Chemotherapy is mainly based on a whole-body treatment with chemotherapeutic agents, and it is inevitable to cause many dangerous side effects associated with the non-selective cytotoxic effect of the medications. However, nanotechnology offers the means to target chemotherapies directly and selectively to cancerous cells and neoplasms, guide in surgical resection of tumors, and enhance the therapeutic efficacy of radiation-based and other current treatment modalities. Therefore, nanotechnology of cancer therapy extends beyond drug delivery into the creation of new therapeutics available only through the use of NM properties. Although small compared to cells, NPs are large enough to encapsulate many small molecule compounds, which can be of multiple types. At the same time, the relatively large surface area of NP can be functionalized with ligands, including small molecules, DNA or RNA strands, peptides, aptamers, or antibodies. These ligands can be used for therapeutic effect or to direct NP fate in vivo. These properties enable combination drug delivery, multi-modality treatment, and combined therapeutic and diagnostic, known as "theragnostic," action.

The physical properties of NPs, such as energy absorption and re-radiation, can also be used to disrupt diseased tissue, as in laser ablation and hyperthermia applications. All of this can add up to a decreased risk to the patient and an increased probability of survival. In addition, the integrated development of innovative NP packages and active pharmaceutical ingredients will also enable the exploration of a wider repertoire of active ingredients, no longer confined to those with acceptable pharmacokinetic or biocompatibility behavior. Immunogenic cargo and surface coatings are being investigated as both adjuvants to NP-mediated and traditional radio- and chemotherapy as well as stand-alone therapies.

9.4.2 Detection and Diagnosis

NPs can selectively target cancer biomarkers and cancer cells, allowing more sensitive diagnosis, early detection requiring a minimal amount of tissue, monitoring of the progress of therapy and tumor burden over time, and destruction of solely the cancer cells.

Thus, nanotechnology can provide rapid and sensitive detection of cancer-related molecules, enabling scientists to detect molecular changes even when they occur only in a small percentage of cells. Thus this nanoscale technology has the potential to generate entirely novel and highly effective therapeutic, diagnostic, or both; and the ability to passively accumulate at the tumor site, to be actively targeted to cancer cells, and to be delivered across traditional biological barriers in the body such as dense stromal tissue of the pancreas or the blood–brain barrier that highly regulates delivery of biomolecules to/from our central nervous system.

9.5   Mechanism of Action

9.5.1   *Targeted Therapy*

Targeted therapy can affect the tissue environment that helps a cancer grow and survive or it can target cells related to cancer growth, like blood vessel cells. Targeting of cancer cells specifically is a vital characteristic of nanocarriers for drug delivery, as it enhances the therapeutic efficacy while protecting normal cells from cytotoxicity. Numerous studies have been carried out to explore the targeting design of NP-based drugs. In order to better address the challenges of tumor targeting and the nanocarrier system design, it is crucial to first understand tumor biology and the interaction between nanocarriers and tumor cells. The targeting mechanisms can be broadly divided into two categories, passive targeting and active targeting (Fig. 9.3).

Passive targeting of NPs is mainly achieved by the EPR effect, which exploits the increased vascular permeability and weakened lymphatic drainage of cancer cells and enables NPs to target cancer cells passively. During passive targeting, the extent and kinetics of NM accumulation at the tumor site are influenced by their size. The nanocarriers need to be smaller than the cut-off of the proportions in the neovasculature, with the extravasation to the tumor acutely affected by the size of the vehicle. Further, the biodistribution of the NM-drug formulation is influenced by blood perfusion, passive interactions with biomolecules along the route, and immunological clearance processes such as phagocytosis or renal clearance.

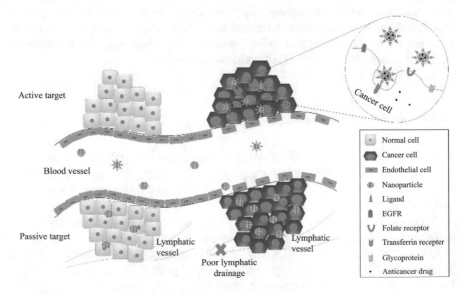

Fig. 9.3   Passive and active targeting of NPs to cancer cells
https://www.frontiersin.org/files/Articles/558493/fmolb-07-00193-HTML/image_m/
fmolb-07-00193-g002.jpg

Active targeting is achieved by the interaction between ligands and receptors. The receptors on cancer cells include transferrin receptors, folate receptors, glycoprotein (such as lectin), and epidermal growth factor receptors (EGFRs). Diverse biomolecules can constitute a ligand, including antibodies, proteins, nucleic acids, peptides, carbohydrates, and small organic molecules such as vitamins. Target substrates can be surface molecules expressed in diseased cells, proteins, sugars, or lipids present in the organs, molecules present (or secreted by tumor cells) in the microenvironment of the diseased cells, or even the physicochemical environment in the vicinity. NM-based smart, targeted systems exploit the multivalent nature of interactions of ligands with the target antigens. When multiple ligand molecules are accumulated onto the nanosystems, there is an overall increase in the avidity of the NPs/NMs for its cognate target. A detailed approach for drug vehicles and targeting has been discussed in greater detail in Chap. 4.

9.5.2 Self-Therapeutic NMs

The application of nanotechnology to cancer therapy could extend beyond drug delivery into the creation of new therapeutics able to destroy the tumors with minimal damage to healthy tissues and organs, as well as the detection and elimination of cancer cells during the initial stage of tumorigenesis. The possible mechanism of action of self-therapeutic NMs in cancer treatment is summarized as under:

Enzyme Inhibition There are some reports in which boron-based compounds (Metalloid Boron) are directly used for the growth inhibition of different cancer cells. The biological roles of boron include the regulation of gene expression, growth, and proliferation. Boron has unique characteristics that allow it to form a covalent bond with carbon. Recently, boron compounds have gained interest as protective and therapeutic agents for prostate cancer and other cancers. Boric acid, the predominant form of boron in plasma, shows boron-mediated anticancer mechanisms in prostate cancer by reducing intracellular calcium signals and calcium storage; decreasing enzymatic activities (serine protease, NAD-dehydrogenases, etc.); and finally inhibiting cancer cell proliferation. The common concept is to arrest cancer growth. Many boron-based NMs such as boron nitride nanosheets, nanotubes, NPs, and rare earth boride nanostructures have been synthesized by several different methods such as Chemical vapor deposition (CVD), hydrothermal methods, and electrochemical methods. Boric acid and boron nitride are the most abundant forms and could be good sources of boron atoms for therapeutic applications.

Oxygen-Capturing Approach The primary tumor or metastasis can grow to a size of about 1–2 mm if provided a sufficient supply of oxygen and nutrients by diffusion, but tumor growth beyond this size requires vascularization, a process called angiogenesis. Some new NMs including copper, carbon, silver, gold, silica, chitosan, and peptides that can be conjugated with antiangiogenic properties have been developed that can not only stop the growth of unwanted blood vessels by blocking them with their soluble byproducts but could also deprive tumor cells of oxygen, resulting in the inhibition of tumor growth. The synthesis of other biocompatible NMs for use as deoxygenation agents based on SiO2 and manganese-based materials is demanding in the current therapeutic systems. Hence, synthesizing these kinds of materials with specific properties such as oxygen capture together with the blockage of extra blood vessels by their byproducts could bring about advances in antiangiogenesis therapy.

Highly Toxic Hydroxyl Radical Ions for Cancer Cell Death This mechanism is utilized in cancer by converting H_2O_2 (present at a much higher concentration than in normal cells, ranging from 100 μM to 1 mM in the tumoral microenvironment) into highly toxic hydroxyl free radicals (OḤ) to efficiently kill cancer cells and suppress tumor growth. For example, using the Fenton chemical process, taking advantage of overproduced H_2O_2 in the tumor, amorphous iron NPs (AFeNPs) are prepared that can be used as cancer theragnostics. The amorphous structure and the ionization of AFeNPs enables the on-demand release of a ferrous ion into the tumor, which leads to the production of an excessive amount of OḤ free radicals in the tumor microenvironment. The excessive amount of OḤ ions so produced can further oxidize protein lipids and DNA to decrease cell viability.

Self-Therapeutic Organic NMs Organic materials such as peptide-based NMs are used in medicine for therapeutic applications. The major advantage of peptide-based therapeutic systems is that they could target cancer cells with lower toxicity to normal tissues. The (KLAKLAK)2 (KLAK) amphiphilic peptide is a well-known antitumor peptide widely used in cancer therapy in vitro and in vivo that suppresses the growth of aggressive tumors by damaging the membrane of mitochondria. The synthesis of these organic NMs always depends on considering their action to inhibit cancer tumors, for example, peptide-based nanoclusters are used to enhance their in vivo stability and penetration abilities inside tumors. Similarly, P-gp inhibitors are conjugated to increase their in vivo stability and enhance their cytotoxicity.

9.6 Drug Release Strategy

It is of great significance to develop precise on-demand drug-delivery systems with high selectivity and responsiveness to enhance anticancer therapy. Stimuli-responsive nanomedicines that can be activated by delicately designed cascade reactions have attracted increasing attention to achieve fine control over the activation process of anticancer agents. To achieve this, stimuli-responsive nanomedicines that can be activated by delicately designed cascade reactions have been developed in recent years. In general, the nanomedicines are triggered by an internal or external stimulus, generating an intermediate stimulus at the tumor site, which can intensify the differences between tumor and normal tissues; the drug release process is then further activated by the intermediate stimulus. There are two categories of nanosystems, open-loop control systems and closed-loop control systems, grouped according to what activation factors stimulate drug release. In open-loop control systems, external factors such as magnetic pulses, thermal pulses, acoustic pulses, or electric fields control drug release. In contrast, in closed-loop systems, the drug release rate is controlled by the presence and intensity of internal stimuli in the vicinity of the target sites. A few current strategies are based on the "chemistry" programmed into the nanosystems that are responsive toward pH or temperature, magnetic field, erosion due to the local chemical environment, redox reaction-based release, and enzyme-mediated release.

9.7 Nanotoxicity

Knowledge of the toxic effects of NMs used for cancer therapy is limited, but it is rapidly growing. Many studies have shown that some NPs demonstrate toxicity in biological systems. Thus research in the internal and external environment is needed; external studies can direct the internal studies. Some researchers have shown that most NPs can release active oxygen and cause oxidative stress and inflammation by the RES (reticuloendothelial system). Acute toxicity resulting from NPs indicate that toxicity depends on the size, coating, and chemical component of the NPs. Also, the systemic effects of NPs have been shown in different organs and tissues. The effects on inflammatory and immunological systems may include oxidative stress or pre-inflammatory cytotoxin activity in the lungs, liver, heart, and brain. The effects on the circulatory system can include prethrombosis effects and paradox effects on heart function. Genotoxicity, carcinogenicity, and teratogenicity may occur as a result of the effects of NPs. Some NPs could pass the blood–brain barrier and cause brain toxicity; of course, more studies are required. It is obvious that most nondegradable NPs are toxic and can influence the body's cells. The biocompatible coatings improve the performance of these NPs, reduce their toxicity, and do not result in negative effects on cells. However, some methods may require modification and some new testing methods may also be needed. It appears that NPs can exacerbate certain preexisting medical conditions and may increase susceptibility to some diseases, which may require modification of testing strategies.

Finally, the toxicity of NPs can differ depending on the experimental method employed. A standardization of toxicity protocols, long-term study of NP toxicity, and the fate of these NMs in human tissue and in the environment need to be further investigated (see also Chap. 3).

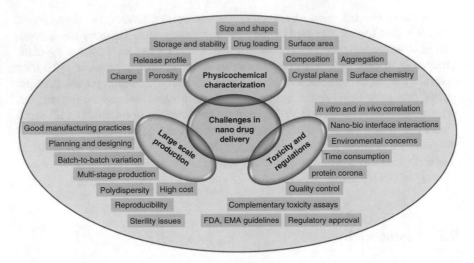

Fig. 9.4 Schematic illustration representing various challenges involved in the delivery of cancer nanotherapeutics
https://media.springernature.com/lw685/springer-static/image/art%3A10.1186%2Fs40
580-019-0193-2/MediaObjects/40580_2019_193_Fig9_HTML.png?as=webp

9.8   Challenges in Nano Cancer Therapy

Nanotechnology applied to cancer therapy has led to a new era of cancer treatment. The use of diverse NMs with desired properties and recent progress in the drug delivery arena have revealed outstanding challenges in cancer therapy and management. A schematic representation of the major challenges in the delivery of cancer nanotherapeutics is depicted in Fig. 9.4. The physicochemical properties of NMs play a significant role in the biocompatibility and toxicity in the biological systems. Therefore, the physicochemical characterization of NPs, their large-scale production, development of toxicity protocols, standardization, and implementation of regulations represent some of the challenges involved in the delivery of cancer nanotherapeutics.

It is anticipated that the NMs will revolutionize the entire health care system based on the dramatic developments made in drug delivery sector over the past few decades. Various types of NPs, including organic and inorganic NPs, have already been widely used in the clinical treatment of several cancer types. These nanocarriers interact with the biomolecules and may tend to aggregate forming a protein corona, disturbing the regular function of nanomedicine formulations, and rendering them ineffective in controlling cancer cell growth. In conjunction with physicochemical properties, the NM storage and stability may also have an influence on their pharmacological performance. However, compared to traditional drugs, NP-based drug delivery systems are associated with improved pharmacokinetics, biocompatibility, tumor targeting, and stability, while simultaneously playing a significant role in reducing systemic toxicity and overcoming drug resistance. These advantages enable NP-based drugs to be widely applied to chemotherapy, targeted therapy, radiotherapy, hyperthermia, and gene therapy. Several studies have revealed the detrimental properties of nanocarriers due to their toxicity. In addition to all the above, a significant setback in nanomedicine commercialization is the clinical translation due to the lack of in-depth understanding of nano-bio interfacial interactions.

Further Reading

Bae KH, Chung HJ, Park TC. Nanomaterials for cancer therapy and imaging. Mol Cells. 2011;31(4):295–302. https://doi.org/10.1007/s10059-011-0051-5.

Gupta PK. Chapter 15: Toxic effects of nanoparticles. In: Toxicology: resource for self study questions. 2nd ed. Seattle: Kinder Direct Publications; 2020a.

Gupta PK. Chapter 14: Toxicology of nanomaterials. In: Problem solving questions in toxicology – a study guide for the board and other examinations. 1st ed. Switzerland: Springer; 2020b.

Gupta PK. Toxic effects of nanoparticles. In: Brain storming questions in toxicology. 1st ed. Boca Raton: Taylor & Francis Group, LLC, CRC Press; 2020c. p. 297–300.

Gupta PK. Fundamentals of nanotoxicology. 1st ed. New York: Elsevier; 2022.

Madamsetty VS, Mukherjee A, Mukherjee S. Recent trends of the bio-inspired nanoparticles in cancer Theranostics review article. Front Pharmacol. 2019 Oct 25. https://doi.org/10.3389/fphar.2019.01264.

Muhammad A, Fahriye D, Vincenzo C, Flavio R. Self-therapeutic nanomaterials for cancer therapy: a review. ACS Appl Nano Mater. 2020;3(6):4962–71. https://doi.org/10.1021/acsanm.0c00762. https://pubs.acs.org/action/showCitFormats?doi=10.1021/acsanm.0c00762&ref=pdf

Navya PN, Kaphle A, Srinivas SP, et al. Current trends and challenges in cancer management and therapy using designer nanomaterials. Nano Converg. 2019;6:23. https://doi.org/10.1186/s40580-019-0193-2.

Zhang Y, Li M, Gao X, et al. Nanotechnology in cancer diagnosis: progress, challenges and opportunities. J Hematol Oncol. 2019;12:137. https://doi.org/10.1186/s13045-019-0833-3.

Chapter 10
Nanomedicine in Immune System Therapy

Abstract With the advent of nanotechnology, the prospects for using engineered nanomaterials (NMs) with diameters of <100 nm in industrial applications, medical imaging, disease diagnoses, drug delivery, cancer treatment, gene therapy, and other areas have progressed rapidly. The potential for NPs in these areas is infinite, with novel applications constantly being explored. The possible toxic health effects of these NPs associated with human exposure are unknown. As a result of acquired cellular damage, NPs can induce different pathways of programmed cell death, including apoptosis, regulated necrosis, and autophagic cell death. Although immunotherapy has made significant advances, the clinical applications of immunotherapy encounter several challenges associated with safety and efficacy. This chapter deals with immunotherapy to improve the ability of immunomodulatory molecules to reach disease tissues, immune cells, or their intracellular compartments, in the context of chronic immune disorders and health problems associated with the use of nanotechnology.

Keywords Nanotechnology · Drug delivery · Immune system · Cancer immunotherapy · Nano-immuno-oncologicals · Nano-immunotherapies · Autoimmune disease · NPs · Immunosuppression · Nanotoxicology · Immunotoxicity · Anti-Inflammatory · Immunostimulation

10.1 Introduction

Nanomedicine in immune system therapy includes a series of colloidal nanoparticles (NPs) to improve the ability of immunomodulatory molecules to reach disease tissues, immune cells, or their intracellular compartments, in the context of chronic immune disorders. A successful response to immunotherapy requires administration of the right immunomodulator dose during the right time at the right tissue, cellular, and even intracellular location. This chapter deals with NPs to improve the ability of immunomodulatory molecules to reach diseased tissues, immune cells, or their intracellular compartments, in the context of chronic immune disorders and adverse effects associated with their use.

10.2 Overview of Immune System

The immune system is the body's defense mechanism against disease and infection and is responsible for targeting and destroying substances that it recognizes as foreign or different from normal, healthy tissues in the body. From a functional perspective, the immune system consists of:

(a) Innate immunity
(b) Adaptive immunity

Although both innate and adaptive immunity are separate, they are interacting and overlapping defensive systems that provide an additional array of defensive weapons. In addition, innate immunity and adaptive immunity are activated by the recognition of molecular shapes that are "foreign" to our body. By distinguishing between "self" and "non-self" these systems are (normally) able to identify, destroy, and remove foreign cells; infectious agents; and large foreign molecules without directly attacking our own cells and tissues.

10.2.1 Innate, or Nonspecific, Immunity

Innate, or nonspecific, immunity is the defense system with which one is born, meaning anything that is identified as foreign or non-self is a target for the innate immune response (natural immunity, and it is so named because it is present at birth and does not have to be learned through exposure to an invader). It thus provides an immediate response to foreign invaders. This system is activated by the presence of antigens and their chemical properties. It protects us against all antigens. Innate immunity involves barriers that keep harmful materials from entering our body. These barriers form the first line of defense in the immune response. Innate immunity, unlike acquired immunity, has no memory of the encounters, does not remember specific foreign antigens, and does not provide any ongoing protection against future infection.

The white blood cells involved in innate immunity are: monocytes (which develop into macrophages), neutrophils, eosinophils, basophils, and natural killer cells. Each one has a different function. Other participants in innate immunity are mast cells; the complement system (the complement system consists of more than 30 proteins that act in a sequence. One protein activates another, which activates another, and so on to defend against infection. This sequence is called the complement cascade); and cytokines (Fig. 10.1). Virtually all cells can contribute to innate immunity by producing certain innate cytokines, particularly the type 1 IFNs, and

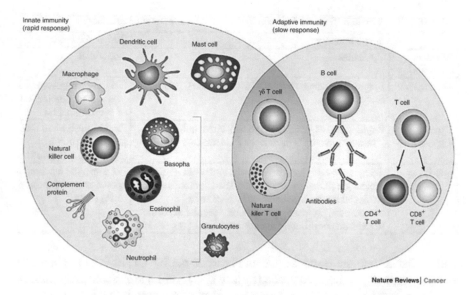

Fig 10.1 Cellular difference between innate and adaptive immunity. (https://microbiologyinfo.com/wp-content/uploads/2016/10/Difference-between-Innate-and-Adaptive-Immunity.jpg)

by responding to these cytokines to induce new and elevated intracellular molecular mechanisms for fighting off infections.

Cytokines are the messengers of the immune system. White blood cells and certain other cells of the immune system produce cytokines when an antigen is detected. Cytokines also participate in acquired immunity. There are many different cytokines, which affect different parts of the immune system:

(a) Some cytokines stimulate activity. They stimulate certain white blood cells to become more effective killers and to attract other white blood cells to a trouble spot.
(b) Other cytokines inhibit activity, helping end an immune response.
(c) Some cytokines, called interferons, interfere with the reproduction (replication) of viruses.

10.2.2 Adaptive Immunity

Adaptive immunity is immunity that occurs after exposure to an antigen either from a pathogen or a vaccination. This part of the immune system is activated when the innate immune response is insufficient to control an infection. A person who recovers from measles, for example, is protected for life against measles by the adaptive immune system, although not against other common viruses, such as those that cause mumps or chickenpox. There are two types of adaptive responses: the

Table 10.1 TThe main differences between innate and adaptive immune memory

	Innate memory	Adaptive memory
Effector molecules	Cytokines	Antibodies
Mechanisms	Epigenetic changes (e.g., DNA methylation, histone acetylation)	Gene rearrangement (somatic recombination of gene segments)
Type of response	Rapid (same as primary response), either enhanced ("trained memory") or reduced ("tolerance")	Rapid (much more than primary response), enhanced/more potent
Specificity	Triggered by any molecule or stressful event (e.g., molecules shared by groups of related microbes or produced by damaged host cells, metabolic compounds, pollutants, etc.), upon a second exposure to the same or different agent/event	For a specific antigen, upon a second exposure to the same

cell-mediated immune response, which is carried out by T cells, and the humoral immune response, which is controlled by activated B cells and antibodies (Fig. 10.1).

The main difference between innate and adaptive memory is that innate memory is non-specific. The main differences between innate and adaptive immune memory are summarized in Table 10.1.

10.3 Applications of NPs

Immunotherapy has emerged as a promising and innovative strategy which is widely used for the treatment of various types of diseases by modulating the host's immune system. NPs show profound immunomodulatory effects (immunosuppressive or immunostimulatory) and, therefore, have been used for the treatment of various types of diseases. For example:

(i) *Immunosuppressive:* Immunosuppressive therapy refers to the downregulation of the immune response which helps in the treatment of various types of auto-immune diseases like type 1 diabetes, obesity, atherosclerosis, and rheumatoid arthritis.

(ii) *Immunostimulatory:* Immunostimulatory therapy refers to that which activates the immune response, thus helping in the treatment of cancer and other infectious diseases. NPs are used for the treatment of cancer and some other infectious diseases. The activation of the host's immunity can be done by various approaches such as the introduction of various cancer vaccines, monoclonal antibodies, immune checkpoint blockers, and cell-based therapies which have been proven to be very effective in many patients. Cancer immunotherapy not only treats cancer by inducing a strong anti-tumor immune response but also controls metastasis as well as prevents its recurrence, hence representing a major advantage over traditional cancer treatments.

An interaction between a NP and the immune system is considered desirable when it may lead to various beneficial medical applications, such as vaccines or therapeutics for inflammatory and autoimmune disorders (Fig. 10.2).

The applications of nano-immuno-oncological therapies are aimed at improving the ability of the patient's own immune system to recognize and kill cancer cells. For example, several categories of NPs enhance the efficacy of:

(i) Cancer vaccines including peptides, adjuvants, or even nucleic acids such as mRNA.
(ii) Immune checkpoint inhibitors or other monoclonal antibodies.

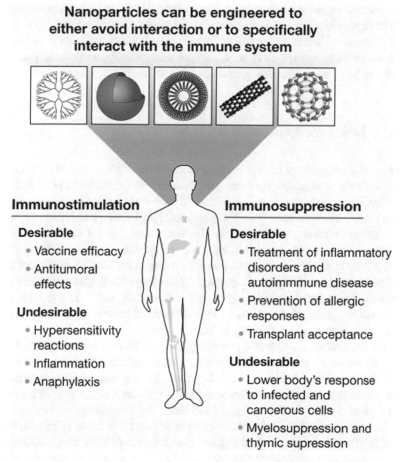

Fig 10.2 Nanoparticle interactions with the immune system. Nanoparticles' effects on the immune cells may benefit treatment of disorders mediated by unwanted immune responses and enhance immune response to weak antigens. On the other hand, undesirable immunostimulation or immunosuppression by Nanoparticles may result in safety concerns and should be minimized. (https://www.ncbi.nlm.nih.gov/pmc/articles/PMC2817614/bin/zee0021051950001.jpg)

(iii) Small drugs or other nano-immunostrategies to reprogram the tumor microenvironment. Nano-immunotherapies for autoimmune diseases include
Alzheimer disease, diabetes, inflammatory bowel disease, systemic lupus ery-
thematosus, multiple sclerosis, and rheumatoid arthritis.

10.4   NPs for Immunotherapy

Immunotherapy is the field of immunology that aims to identify treatments for diseases through induction, enhancement, or suppression of an immune response. This
perspective will focus on two areas of immunotherapy, activating immunotherapies
for cancer and suppressive immunotherapies for autoimmunity both of which have
seen a resurgence in interest in recent years and are likely to transform the treatment
of many human diseases in the future. Immunomodulators are natural or synthetic
molecules that can normalize or modulate our body's immune system. These substances can be either immunosuppressants or immunostimulants.

10.4.1   Immunosuppression

Many novel strategies for immunosuppression rely on NPs as delivery vehicles for
small-molecule immunosuppressive compounds. As a consequence, efforts in
understanding the mechanisms by which NPs directly interact with the immune
system have been overshadowed. The immunological activity of NPs is dependent
on the physiochemical properties of the NPs and its subsequent cellular internalization. Immunosuppressive strategies that pin-point specific tissue or pathways, while
reducing systemic side effects, are a potential panacea for treating autoimmune disorders and complementing pharmaceuticals. To achieve these goals, many chemical
and biomedical researchers have engineered NPs to carry and locally deliver immunosuppressive agents. This can be considered "indirect immunosuppression," where
the NP solely serves as the delivery vehicle.

   As an alternative, a small but prevailing body of literature is reassessing direct
immunosuppression by NPs to be exploited as a complement for drug therapies or
organ/tissue transplantation that otherwise would be rendered ineffective or rejected
by the native immune response. In general, the physiochemical properties of NPs
are important factors that significantly influence the interaction of NPs and cells.
Gold NPs are particularly exemplary systems to illustrate these effects. Spherical
gold NPs between 5 and 30 nm in diameter are capable of interacting with cells by
passive means; however, larger NPs and rod-like NPs are more commonly internalized via complex uptake processes. The surface-coating of the gold NPs also effect
cellular uptake. Where small-molecule organic ligands like citrate or lipids may
promote stability and passive cellular uptake, macromolecular coatings like poly
(ethylene glycol) may result in protein adsorption and reduction in cellular uptake.

Accordingly, the wide variation in size, shape, and surface coating of NPs limits broad generalizations about the interactions of NPs with the immune system.

A plethora of therapeutic components, adjuvants, and immunomodulatory compounds can be effectively delivered by NP-based platforms, including protein antigens, DNA, and RNA. The approach of nanovaccine stimulation of the antitumor immune response is shown in Fig. 10.3.

Nanoparticles that can directly interact with the immune system may be categorized as:

(a) Metal NPs
(b) Metal-oxide NPs
(c) Carbon NMs
(d) Polymer NPs and macromolecules

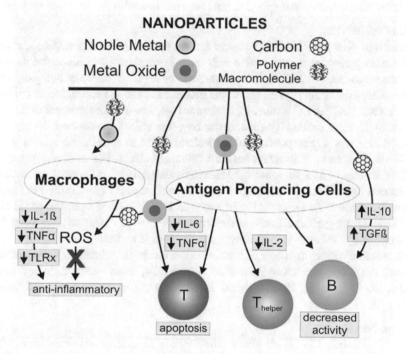

Fig 10.3 Direct interactions of NPs with the immune system. Dependent on physiochemical properties, the NPs interact with the constituents of the immune response, including macrophages, antigen presenting cells, B cells or T cells. Direct effects of carbon nanomaterials include the upregulation of transforming growth factor-β (TGFβ), interleukin-10 (IL-10), decreased B cell activity, as well as apoptosis. Metal-oxide NPs can directly affect adaptive immune cells, and materials like cerium oxide NPs scavenge reactive oxygen species (ROS) and operate as anti-inflammatory agents. Polymer NPs and macromolecules (dendrimers) exhibit an array of immunosuppressive effects. The pathways shown are representative examples by which different nanoscale products might suppress the immune system. (https://www.ncbi.nlm.nih.gov/pmc/articles/PMC4950368/bin/10.1177_1535370216650053-fig1.jpg)

A few of the NPs that can be used for immunosuppression as delivery vehicles for small-molecule immunosuppressive compounds are briefly discussed as under:

Metal NPs

Noble metal NPs, such as gold and silver, interact with both the innate and adaptive immune systems, but few reports uncover the mechanism behind noble metal NPs' ability to elicit an immunosuppressive response. Injection of organo-gold compounds has been utilized for nearly a century to treat inflammation, and only recent reports provide the first analysis of the biochemical pathway in which gold NPs may reduce inflammation. Citrate-coated gold NPs showed anti-inflammatory activity and inhibited cellular responses induced by interleukin 1 beta (IL-1β). IL-1β is an inflammatory cytokine that acts as an arbiter between the innate and adaptive immune response; moreover, common inflammatory disorders, for example rheumatoid arthritis, are mediated by IL-1β production. The studies on the immunotoxicity of Ag NPs are very limited. Ag NPs induced reactive oxygen species (ROS) and inflammation, indicating its potential interference with the immune system.

Metal-Oxide NPs

Metal-oxide NPs have been reported to induce both immunosuppressive and anti-inflammatory responses. Iron oxide NPs have been shown to reduce the humoral immune response. The humoral response involves the recognition of antigens, allergens, pathogens, or foreign bodies in the blood. Free-radical formation in the body has been attributed to inflammation, tissue damage, and the development of diseases (Chap. 3). Cerium oxide NPs (nanoceria) have the ability to reduce ROS and may be used as a novel therapeutic tool for inflammation treatment. Nanoceria with a small diameter has a significant anti-inflammatory effect. For example, nanoceria with a diameter of 3–5 nm scavenged free radicals inhibited inflammatory mediator production in J774A.1, the murine macrophages. These NPs can mimic the superoxide dismutase active scavenging of superoxide (O_2^-). Unlike cerium oxide NPs, the single defining characteristic of the immunological response to zinc oxide NPs is cytotoxicity related to oxidative stress from the increased production of ROS. Another titanium oxide NP, unlike zinc oxide, has been reported as noncytotoxic in vitro, but it can induce an immunosuppressive response. In a tumor model, titanium oxide NPs inhibited T-cells, B-cells, macrophages, and natural killer cells.

Carbon Nanotubes

Carbon nanotubes (CNTs) have been introduced in pharmacy and medicine for drug delivery systems in therapeutics. CNTs exhibit strong free-radical scavenger properties and can be utilized as anti-inflammatory agents. Similar to cerium oxide NPs, fullerenes (C_{60}) decrease the level of ROS by efficiently scavenging free radicals. Unlike cerium oxide NPs, fullerenes distribute the free radicals through their aromatic structure. Fullerenes reduce ROS of both hydroxyl and superoxide radicals, and many of the radical scavenging properties are moderated by the functionalization of fullerenes with water-soluble ligands. After inhalation exposure, CNTs can induce systemic immunosuppression in mice, including the production of prostaglandin and IL-10 and T cell dysfunction. Inhalation of CNTs results in non-monotonic systemic immunosuppression (reduced T-cell-dependent antibody against sheep erythrocytes and T-cell proliferative ability and decreased natural killer cell activity). This suppression is accompanied by increased spleen gene expression of IL-10, which is an anti-inflammatory cytokine, and NAD(P)H oxidoreductase.

In recent years, fullerene NPs have received extensive attention due to their unique physical and chemical properties. Properly modified fullerene NPs have excellent biocompatibility and significant anti-tumor activity, which makes them have broad application prospects in the field of cancer therapy.

Quantum Dots
Quantum Dots (QDs) can induce the generation of ROS by transferring energy to nearby oxygen molecules. In vitro studies have shown that QDs can induce the production of ROS and lead to multiple organelle damage and cell death. Using a mouse model of multiple sclerosis (MS), it is shown for the first time that QDs can be used to generate immunological tolerance by controlling the density of self-antigen on QDs.

Polymeric NPs
Polymeric NPs based on polylactide (PLA) or poly(lactide-co-glycolide) (PLGA) have attracted much attention as delivery vehicles. For example, cyclosporine A (CsA) has excellent biocompatibility, and this nanoparticulate CsA delivery technology constitutes a strong basis for future targeted delivery of immunosuppressive drugs with improved efficiency and potentially reduced toxicity. In addition, polystyrene particles have the potential to halt the disease process in autoimmunity. Antigen-decorated polystyrene particles with a diameter of 500 nm can induce T-cell tolerance and ameliorate experimental autoimmune encephalomyelitis by inactivating pathogenic T cells.

10.4.1.1   Immunostimulation

Immunostimulatory therapy refers to that which activates the immune response, thus helping in the treatment of cancer and other infectious diseases. Engineered NPs can specifically be designed to either target or avoid interactions with the immune system. An interaction between a NP and the immune system is considered desirable when it may lead to various beneficial medical applications, such as vaccines or therapeutics for inflammatory and autoimmune disorders. NP immunogenicity is drawing interest because NPs have been shown to improve the antigenicity of conjugated weak antigens and thus serve as adjuvants (some NPs have been shown to be antigenic themselves). This property has been shown to depend on particle size and surface charge and can significantly contribute to the development of improved vaccine formulations. Particle size has been reported as a major factor in determining whether antigens loaded into NPs induce type I (interferon-γ) or type II (IL-4) cytokines, thereby contributing to the type of immune response. A leading hypothesis on why nanotechnology-driven formulations are effective in vaccine development is that non-soluble NPs provide controlled, slow release of antigens, creating a depot at the site of injection and providing protection in the destabilizing in vivo environment. Several studies have reported cytokine induction by different types of NMs (gold colloids, dendrimers, polymers, lipid NPs, etc.). NP size has been suggested as a leading parameter that determines a NP's potential to induce cytokine responses. However, a few studies have shown that cytokines are induced not by NPs per se but by surfactants or bacterial endotoxins present in the formulation. The immunostimulatory effects of engineered NPs that have been and are being widely used in immunotherapy are summarized as under.

Metal NPs

Gold NPs (GNPs) are also used in immunotherapy due to their low cytotoxicity, tunable surface chemistry, and easily controllable shape and size. GNPs are an important class of immunostimulatory NPs which show its response by activating macrophages and their subsequent differentiation into dendritic-like cells, leading to T-cell proliferation and cytokine release. GNPs are also found to be useful as an adjuvant for antibody production in mice, and its immunogenic property can be further increased if used in combination with another immunostimulant, cytosine-phosphate-guanine oligodeoxynucleotides (CpG-ODNs).

Silica NPs

Silica NPs are mostly biocompatible and widely used in various biomedical applications, like bioimaging, tumor targeting, and drug/vaccine delivery. For example, MSN (mesoporous silica NP) is an efficient antigen delivery vehicle that can elicit both humoral and cell-mediated immune responses without causing any cytotoxic effect, having self-adjuvant potential and biocompatibility in vaccine delivery applications as well as its self-adjuvant effect by immunizing.

Magnetic NPs

Magnetic NPs have been widely used in the theranostic domain due to their magnetic resonance imaging (MRI) properties. Among these NPs, superparamagnetic iron oxide NPs (SPIONs) are attracting much interest from many researchers for cancer theranostic applications. These NPs are often coated with a layer of biocompatible materials in order to reduce their aggregation.

Carbon Nanotubes

Carbon nanotubes have many interesting properties. In particular, their photohyperthermic effect by near-infrared (NIR) irradiation can be used to kill cancer cells, thus it can be used as photohyperthermic therapy. By increasing the solubility, CNTs can be used as an immunostimulatory agent as well as a delivery vehicle for antigens and adjuvants for cancer immunotherapy.

Polymeric NPs

Polymeric NPs are the most widely used immunostimulatory NPs as they exhibit excellent biocompatibility, biodegradability, chemical stability, water solubility, and high capacity to load immune-related components. The commonly used polymeric

NPs in cancer immunotherapy are poly (D, L-lactic-coglycolic acid) (PLGA), poly (g-glutamic acid) (PGA), poly (D,L-lactide-co-glycolide) (PLG), poly (ethylene glycol) (PEG), poly ethylenimine (PEI), and chitosan NPs. These NPs have extensively been employed as an effective immunostimulatory adjuvant in vaccination. Chemotherapeutic PLGA formulations with varying properties (i.e., shape, carrier, size, etc.) are currently available and FDA-approved for several types of cancer treatments.

Protein NPs have also been used as effective vaccine platforms for delivering tumor antigens and adjuvants to induce a strong anti-tumor immune response. For example, biomimetic protein NPs could effectively co-deliver peptide epitopes and CpG oligodeoxynucletides (CpG ODN) activator to dendritic cells (DCs), resulting in increased and prolonged $CD8^+$ T cell activation (often called cytotoxic T lymphocytes, or CTLs) as well as enhanced antigen cross-presentation. Micelles have also been used as an effective nanocarrier of antigen/adjuvant for enhancing the potency of cancer vaccines.

10.5 Targeted Delivery Systems

With the development of nano- and micro-particles, there has been a growing number of immunotherapy delivery systems developed to elicit innate and adaptive immune responses to eradicate cancer cells. This can be accomplished by training resident immune cells to recognize and eliminate cells with tumor-associated antigens or by providing external stimuli to enhance tumor cell apoptosis in the immunosuppressive tumor microenvironment. Targeted delivery systems include any delivery platform with a targeting moiety such as antibodies, peptides, and glycoprotein to improve the efficiency and specificity of drug delivery. Novel strategies, especially improved delivery strategies, can more effectively target tumors and/or immune cells of interest, increase the enrichment of immunotherapies within the lesion, and reduce off-target effects. Some materials, such as lipids, polymers, and metals, have been used to exploit delivery strategies. At present, new delivery strategies are being researched and developed for immunotherapy, including NPs, scaffolds, and hydrogels. These delivery platforms offer many advantages for immunotherapy compared to separate therapeutic agents.

A targeted NP-based gene delivery concept was translated from bench to bedside for the first time with Rexin-G and Reximmune-C formulations. Rexin-G and Reximmune-C are targeted multilamellar vesicle-based NPs approximately 100 nm in size. The former encapsulates a construct encoding a dominant-negative mutant of human cyclin G1 protein; the latter carries a cytokine, granulocyte-macrophage colony-stimulating factor. These two drug therapies exemplify the two-tier complementary approach aimed at both tumor eradication and cancer vaccination. The first step involves administration of Rexin-G to target and kill the tumor cell in a programmed fashion, whereas the second step employs Reximmune-C to induce a localized cancer autoimmunization.

10.6   Nanotoxicity and Biosafety Issues

file:///C:/Users/P.K.%20Gupta/Downloads/nanomaterials-10-02186.pdf.

Nanomedicine is a rapidly growing field that uses nanomaterials for the diagnosis, treatment, and prevention of various diseases such as cancer and immunotherapy. Various biocompatible nanoplatforms with diversified capabilities for tumor targeting, imaging, and therapy have materialized to yield individualized therapy. The major roadblock is that we are yet to fully understand the toxicity profiles of NP-mediated immune response. It is unclear whether the codelivery of immunotherapeutic agents with chemoradiotherapy will result in intolerable side effects. Therefore, it is mandatory to study the toxicity profiles of NPs and its modulation with various components of nanoformulations. The increasing understanding of nanotoxicity paradigms has recently resulted in important benchmarks for the safe design of nanomaterial-based drug delivery systems aiming to fight various medical problems including cancer and immune therapy. However, due to their unique properties brought about by their small size, safety concerns have emerged as their physicochemical properties can lead to altered pharmacokinetics, with the potential to cross biological barriers. In addition, the intrinsic toxicity of some of the inorganic materials (i.e., heavy metals) and their ability to accumulate and persist in the human body has been a challenge to their translation. A small subset of NPs that are currently undergoing preclinical investigation, for example, carbon- and metal-based NPs, typically display cytotoxic properties. The harmful effects of these particles are mainly based on ROS generation, disruption of cellular compartments, and immune reactions (Chap. 3). Nevertheless, the inherent toxic properties of such NPs could be exploited to ablate diseased tissue, as long as healthy organs are protected through selective targeting. Moreover, there are several strategies (e.g., surface modification) that can be utilized to eliminate toxicity. Successful clinical translation of these NPs is heavily dependent on their stability, circulation time, access and bioavailability to disease sites, and their safety profile.

From proof-of-concept to commercialization, our increasing understanding of drug delivery has dramatically promoted the commercial transition of nanomaterial-based drug delivery systems into the market for improving our well-being. After the success of the first generation of nanomedicines, the next generation of nanomedicines with improved site-targeting, stimuli-responsive drug release, and multimodal capacities are currently undergoing clinical trials. However, the biosafety issue and the complex of bio-nano interactions in the human body could hamper the development of nanomedicines, and thus will continually gain interest from both researchers and the pharmaceutical business in the future. Indeed, it has even been suggested that a good number of studies have arrived at erroneous conclusions, due in part to the fact that excessively high doses of NMs have been applied, that is, doses that are unrealistic from the point of view of occupational or environmental exposure. Meanwhile, we are faced with an onslaught of new nanomaterials, and it is becoming more and more apparent that we cannot resort to traditional risk assessment approaches with their strong reliance on animal data because NPs are highly heterogeneous, with very diverse combinations of chemical composition, core-shell structure, shape, and functionalization. This poses a challenge in the experimental assessment of the relationship between their physicochemical properties and their toxicological effects.

Overall, the effects of NP toxicity on the immune system are complex and multifaceted as it involves a variety of different cell types across organ systems. In general, the current research suggests that Au NPs present limited toxicity to the cells of the immune system including macrophages, lymphocytes, and dendritic cells. However, the results with IO NPs are more mixed, with some functional impairment effects demonstrated, though the cytotoxic potential to immune cells remains low. Additionally, the surface coating and electrostatic charge of the NPs play a key role in the immunotoxicity profile and potential evasion of the immune system responses in targeted drug-delivery systems, an area for additional research.

10.7   The Future Prospective in Immunotherapy

Nanotechnology is the engineering of functional systems at the molecular level. The field combines elements of physics and molecular chemistry with engineering to take advantage of unique properties that occur at nanoscale. Immunotherapy is now widely considered a landmark therapeutic strategy that could radically alter conventional therapy. However, further understanding of therapeutic effect prediction and resistance mechanisms for immunotherapy is necessary. Immunotherapy, including immune checkpoint inhibition, cannot provide any anti-tumor effect unless T-cells interact with the cancer-associated peptide and the major histocompatibility complex (MHC) molecules, thus recognizing cancer cells. Recent breakthroughs in the use of checkpoint inhibition, when combined with cancer vaccination, will make this feasible: The key factor is to target the relevant cancer antigen. For autoimmune diseases, one is dependent on nonspecific immunosuppressive drugs for far too long. We have failed to learn from the allergy field where effective immunotherapy is achieved by targeted desensitization using allergy-associated antigens. The antigen-specific immunotherapies are the new era of immunotherapy for autoimmune diseases where again the key factor is to target the relevant antigen, in this case, the self-antigen. Today, advances in next-generation sequencing and bioinformatics have made it possible to catalog all genomic mutations in individual patients. Currently, individualized vaccine therapy targeting neo-antigens, which are gene mutation antigens, in several solid tumors, including hepatocellular carcinoma (HCC), are being developed. Clinical trials are already underway for individualized vaccine therapy in Europe, North America, and China in several cancers, and the preliminary results are beginning to be reported. Meanwhile, immunotherapy has the possibility of causing immune-related adverse events different from conventional therapy, and more severe management may be required.

The growing interest in the future medical applications of nanotechnology is leading to the emergence of new applications in various disease conditions including cancer and immunotherapy. Scientists are now hunting for biomarkers to better understand who will and who won't respond to immunotherapy treatments. These can be genetic mutations or proteins from tissue or blood that will help them figure out which patients will benefit most from immunotherapies. Some cancer treatments target single genetic mutations, but immunotherapy biomarkers are more complex and could involve a number of genes and proteins. It is expected that further development of novel immunotherapy can have innovative benefits for many patients suffering from various immune system–related diseases.

10.8 Challenges and Opportunities

Immunotherapy as a whole is rapidly developing. However, the delivery technology for immunotherapy is still in its infancy. Novel delivery strategies that improve immunotherapy are introduced for controlled release, local delivery, and increased stability. Many of the delivery technologies described not only provide a way for improving immunotherapy but also to overcome the inherent heterogeneity of cancer. Although immunotherapy has made significant advances, the clinical applications of immunotherapy encounter several challenges associated with safety and efficacy. For example, in terms of safety, immunotherapy can cause fatal adverse effects in some patients, including autoimmune reactions, cytokine release syndrome, and vascular leak syndrome. Novel strategies, especially improved delivery strategies, are able to more effectively target tumors and/or immune cells of interest, increase the enrichment of immunotherapies within the lesion, and reduce off-target effects. Some materials, such as lipids, polymers, and metals, have been used to exploit delivery strategies. At present, new delivery strategies are being researched and developed for immunotherapy, including NPs, scaffolds, and hydrogels. These delivery platforms offer many advantages for immunotherapy compared to separate therapeutic agents. Another important issue is the development and synthesis of NMs, which are the foundations for the development of cancer immunotherapy delivery strategies and require greater resource acquisition and cost reduction. Producing large-scale industrial samples at a cost that is affordable to patients is a challenge, especially in the early stages. Therefore, several design guidelines, including treatment stability, scalability, and cost and complexity of production, are fundamental issues to consider for clinical translation.

Further Reading

Akazawa Y, Suzuki T, Yoshikawa T, Mizuno S, Nakamoto Y, Nakatsura T. Prospects for immunotherapy as a novel therapeutic strategy against hepatocellular carcinoma. World J Meta Anal. 2019;7(3):80–95. https://doi.org/10.13105/wjma.v7.i3.80.

Anfray F, Mainini F, Torres A. Nanoparticles for immunotherapy. In: Parak WJ, Feliu N, editors. Frontiers of nanoscience, vol. 16. Elsevier; 2020. p. 265–306.

Batty CJ, Tiet P, Bachelder EM, Ainslie KM. Drug delivery for cancer immunotherapy and vaccines. Pharm Nanotechnol. 2018;6(4):232–44. https://doi.org/10.2174/2211738506666180918122337. PMID: 30227827; PMCID: PMC6534808.

Damasco JA, Ravi S, Perez JD, Daniel E, Hagaman DE, Melancon MP. Understanding nanoparticle toxicity to direct a safe-by-design approach in cancer nanomedicine – review. Nano. 2020;10:2186. https://doi.org/10.3390/nano10112186.

Gupta PK. Toxic effects of nanoparticles. In: Toxicology: resource for self study questions. 2nd ed (Chapter 15). Kinder Direct Publications; 2020a.

Gupta PK. Toxicology of nanomaterial particles. In: Problem solving questions in toxicology - a study guide for the board and other examinations. Ist ed, Chapter 14. Switzerland: Springer Nature; 2020b.

Gupta PK. Toxic effects of nanoparticles. In: Brain storming questions in toxicology. Ist ed. Taylor & Francis Group, LLC. CRC; 2020c. p. 297–300.

Gupta PK. Fundamentals of Nanotoxicology. 1st ed. USA: Elsevier Inc.; 2022.

Jiao Q, Li L, Qingxin M, Zhang Q. Immunomodulation of nanoparticles in nanomedicine applications. Bio Med Res Int. 2014, 2014., 426028:1–19. https://doi.org/10.1155/2014/426028.

Ngobili TA, Daniele MA. Nanoparticles and direct immunosuppression. Exp Biol Med (Maywood, N.J.). 2016;241(10):1064–73. https://doi.org/10.1177/1535370216650053.

Patil M, Mehta DS, Guvva S. Future impact of nanotechnology on medicine and dentistry. J Indian Soc Periodontol. 2008;12(2):34–40. https://doi.org/10.4103/0972-124X.44088.

Thakur N, Thakur S, Chatterjee S, Das J, Sil PC. Nanoparticles as smart carriers for enhanced cancer immunotherapy. Front Chem. 2020;8:597806. https://doi.org/10.3389/fchem.2020.597806.

"Nanotoxicology of NanbioMedicine".

NanobioMedicine is emerging as a potential solution for many medical science problems and will change the face of diagnostics, imaging, tissue grafting, therapeutics, and drug delivery, cancer therapy, immune system therapy, tissue engineering and regenerative medicine, dentistry, and so forth in the very near future. The information provided is very concise and crisp for easy grasp of the material.

Thus, the book will be extremely useful for scientists and individuals from academia, industry, and government from a broad field including student of medicine, dentistry, veterinary, biology, toxicology, agricultural scientists, food packaging industrialists, nanotechnologists, as well as nanomaterial scientists and individuals with interest in nanotoxicology and nanotechnology.

Index

© The Editor(s) (if applicable) and The Author(s), under exclusive license to
Springer Nature Switzerland AG 2023
P. Gupta, *Nanotoxicology in Nanobiomedicine*,
https://doi.org/10.1007/978-3-031-24287-8

177

Printed in the United States
by Baker & Taylor Publisher Services